Audio/Video Cable
Installer's Pocket Guide

D1011755

Audio/Video Cable Installer's Pocket Guide

Stephen H. Lampen

McGraw-Hill

New York   Chicago   San Francisco   Lisbon
London   Madrid   Mexico City   Milan
New Delhi   San Juan   Seoul
Singapore   Sydney   Toronto

McGraw-Hill

A Division of The McGraw-Hill Companies

2 3 4 5 6 7 8 9 0 DOC/DOC 0 7 6 5 4 3

ISBN 0-07-138621-1

The sponsoring editor for this book was Stephen S. Chapman and the production supervisor was Sherri Souffrance. It was set in New Century Schoolbook by Patricia Wallenburg.

Printed and bound by R. R. Donnelley & Sons Company.

This book was printed on recycled, acid-free paper containing a minimum of 50% recycled, de-inked fiber.

McGraw-Hill books are available at special quantity discounts to use as premiums and sales promotions, or for use in corporate training programs. For more information, please write to the Director of Special Sales, Professional Publishing, McGraw-Hill, Two Penn Plaza, New York, NY 10121-2298. Or contact your local bookstore.

Preface

This book is a companion volume to *Wire, Cable, and Fiber Optics for Video and Audio Engineers* (McGraw-Hill, 1995). Whereas that book focused on the design, construction, and performance of cables, this book focuses on the application and installation of those cables. In addition, this book is specifically addresses cables intended for audio and video applications. These applications include many new options such as the use of Category cables to run audio or video signals and the emerging networked architecture of audio and video signals.

Whereas the first book was intended for engineers, as indicated by the title, this book is intended for a much wider range of "installers," from the professional broadcast installer to the home theater enthusiast. For those with the least exposure to wire and cable, this books begins with a couple of "basic" chapters. Wire and cable are not "brain surgery" and are no more complicated than any other manufactured product.

Because technology is ever-changing, I hope that readers will feel free to comment on the contents of this book. You can contact me at shlampen@aol.com.

Steve Lampen
San Francisco, California
January 2002

Acknowledgments

This book would not have been possible without the help and support of a number of people. First is Steve Chapman, at McGraw-Hill, whose tireless advocacy of this book made it possible in the first place.

Portions of this book originally appeared in my column, "Wired for Sound" in *Radio World Magazine*. Many thanks to my editor Paul McLane for his help and encouragement.

Then there are the people whose input into the final draft made it readable and accurate: Margie Friedman of Zack Electronics, Dane Ericksen of Hammett & Edison, Art Lebermann of KGO Radio, Michael Masucci of Belden Electronics Division, Terry McGovern of Lucasarts, Dale Reed of Trompeter Electronics, and Bill Ruck, broadcast engineer.

Lastly, I acknowledge the support of my wife, Debra, who put up with a lot of late nights and shepherded this project to its final goal.

The Basics

Conductors

Wire and cable

What is the difference between wire and cable? Wire is a single insulated conducting element and cable is a group of two or more insulated conducting elements. Although there are nonmetallic and even fiber optic (glass) cables, we will focus on the metallic variety.

This book will concentrate on the following six types of wire and cable:

1. The single insulated conductor wire, also known as hook-up or leadwire, is used internally in equipment to hook up components. Larger sizes are used externally to deliver power or as ground conductors (Figure 1-1).

Figure 1-1 Single conductor.

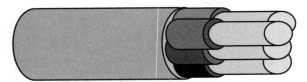

Figure 1-2   Multiconductor.

2. The multiconductor cable, which uses two or more insulated single wires, often carries control signals such as
   switches and lights (Figure 1-2).

3. A twisted pair consists of two insulated wires twisted
   together. Most professional audio signals run on twisted pairs. Very high-quality twisted pairs are used for
   computer signals (Figure 1-3).

4. The multipair "snake" cable consists of two or more
   individual twisted pairs under a common jacket. These
   are used in analog or digital audio to carry multiple
   channels of audio (Figure 1-4).

5. The coaxial cable, or "coax," has one conductor inside
   the other conductor but both share a common axis. This
   type of cable is used to carry many different video signals including analog or digital video, CATV/broad-
   band, or satellite. A lower-quality version is the standard interconnect cable for consumer audio equipment
   (Figure 1-5).

Figure 1-3   Twisted pair.

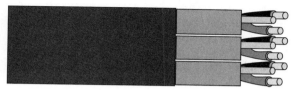

Figure 1-4 Multipair snake cable.

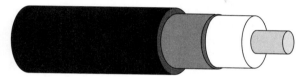

Figure 1-5 Coax.

6. Multicoax cables are made up of two or more coaxial cables. These cables are used to carry multiple video signals. More commonly, they carry the constituent parts of a video signal (component video). The cables are color coded red, green, and blue for that application, and for that reason, are often called RGB cables (Figure 1-6).

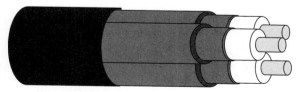

Figure 1-6 Multicoax.

Many of these cables are used for similar applications. For instance, coaxial cables and twisted pair cables are used to carry analog or digital audio, and both have advantages and disadvantages.

Metals

Conductors are materials that conduct electricity. Although there are a few nonmetallic conductors, most common conductors are made of metals. Each type of metal has different physical and electrical properties and different performance. Table 1-1 lists metals commonly used for wire and cable. These are listed by their *direct current* (DC) resistances in circular milliohms per foot at a temperature of 20°C.

TABLE 1-1   Metals and Resistance

Metal	DC Resistance
Silver	9.9
Copper	10.4
Gold	14.7
Aluminum	17
Nickel	47
Tin	69.3
Steel	74

For more on resistance, see page 37.

Oxidation and Corrosion

Metals react to what is around them. If conductors are bare, there is air around them. Moisture and chemicals in the air, especially salts and sulfur, promote chemical reactions. If plastic surrounds the wire, then the wire is protected from the atmosphere. Of course, at the ends of the cable, this protection is removed.

Chemicals in the plastics extruded (melted) around wires can affect the wire itself. This is the reason older cable designs, often military specifications, sometimes called for "noncontaminating" jackets. The chemicals used to make these jacket do not separate or affect the other plastics or the metals.

There are three reactions—oxidation, corrosion, and electrolysis—that can affect the performance of a cable.

Oxidation

Metals can combine with oxygen in the air to produce oxides (Table 1-2). Coating or insulating wires can prevent this effect.

TABLE 1-2   Oxidation

Metal	Oxide	Conductivity
Silver	Silver	Same as silver
Copper	Cupric	Semiconductor
Gold	None	
Aluminum	Aluminum	Nonconductor
Steel	Iron	High resistance

The time it takes for a wire to oxidize depends on whether the cable is indoors or outdoors and the quality and precision of the insulation surrounding the cable. The plastics put on wire sometimes can accelerate oxidation. See Compound Migration on page 34.

Oxidation on bare wire can take anywhere from seconds, if exposed outside, to never, if well protected indoors. It is not uncommon to find house wiring 100 or more years old in excellent condition, with no visible oxidation.

Corrosion

Corrosion is a chemical reaction between dissimilar metals. Table 1-3 shows the potential differences for various metals. All you need is an electrolyte between them to set up the reaction. The electrolyte can be as ordinary as rainwater, which contains salts or sulfur from the atmosphere. Metal can be eaten away or deposited.

TABLE 1-3   Metals and Reaction Potentials

Metal	Potential (volts)
Gold	+1.692
Silver	+0.799
Copper	+0.521
Hydrogen	0 (reference)
Lead	−0.1262
Tin	−0.1375
Nickel	−0.257
Aluminum	−1.662

THE FIRST WIRE

There is evidence that the first wire ever used to carry electricity was made by the Parthians, a people who lived near the Caspian Sea around 200 BC. They used this wire to plate jewelry, which required metal conductors. Gold plated jewelry from their culture still exists in many museums. To plate a metal object, suspend it in an electrolyte (grape juice is thought to be what they used) with a bar of gold. Just attach a wire to each and run an electrical current through, which forces the gold to bind to the surface of the piece to be plated. The process is called "electrolysis." The final proof was earthenware jars, discovered by Greek archaeologist Wilhelm Konig in 1938. These jars contained a copper cylinder and had an iron rod down the center. Fill one with grape juice and you have a 2-volt battery!

If dissimilar metals are put in a cable, they should be separated in some way, or chosen to minimize the potential difference. For example, bare copper braid over aluminum foil shielding is not recommended, because the potential is almost 2 volts. A tinned-copper braid over aluminum foil is preferable. The potential is now reduced to just over 1.5 volts.

Corrosion becomes a serious consideration with cables installed outside in direct-burial or high moisture environments. Cables for these applications are specially designed to withstand the rigors of outdoor installation and moisture. Most often, this calls for a tinned-copper solid conductor inside high-density polyethylene, which is one of the most rugged water-resistant jacket compounds, combined with carbon black pigment for *ultraviolet light* (UV) resistance. See Jackets on page 34.

These interactions should also be considered when choosing connectors, but that is beyond the scope of this book. Contact the connector manufacturer for more information on the metals used in their connectors and the corrosion potential when crimped, soldered, or otherwise connected to a cable. Lead is included in Table 1-3 because it is commonly used in the process of soldering wires to connectors or to each other.

Silver

Silver is the most conductive metal, but it is also expensive and hard to work with. It cannot be annealed so it remains brittle (see "Annealing" on page 9). However, when silver corrodes, the silver oxide that forms on the outside of the wire has exactly the same conductivity as the silver underneath. Adding other metals to decrease the brittleness also increases the resistance. This is the reason you sometimes see silver-coated or "silver-clad" copper or variations with silver coating. There are other reasons to use silver coatings, such as the *skin effect* (page 47). When high-temperature plastics are extruded, such as certain types of Teflon, silver-clad copper helps protect the copper conductor and prevent oxidation.

Connectors often are designed to be "gas tight" to minimize corrosion. These include *insulation-displacement connectors* (IDC) that automatically cut through, but do not remove, the insulation on each wire. This protects the actual point of contact from corrosion. IDCs are common in data connectors, such as the RJ-45, multipair audio punch-down connection blocks, punch-down patch panel connection points, and even some "no tools needed" XLR® 3-pin connectors.

Copper

Copper is the most commonly used conductor for communications cables. Copper is easy to work with. It can be annealed to restore its natural flex-life (see insert for more details on annealing). It has only slightly more resistance than solid silver and is much more reasonably priced.

Copper has one serious drawback. When copper oxidizes, the resulting material, copper oxide, is a semiconductor, which can affect high-frequency signals that flow on the surface of the wire. For low frequencies, which use the entire wire as a conductor, cupric oxide is less of a concern. See Skin Effect on page 47.

> **ANNEALING**
>
> Annealing is a process used to restore the flexibility and flex-life ("flexes until failure") of a wire. To make a wire of a given gage, a larger wire is drawn through a "die." After being drawn a number of times, a copper wire becomes brittle and is easily broken when flexed. So this wire, after drawing, is put into an "annealing oven." At 700°F to 1200°F, this oven is not hot enough to melt the copper but is hot enough to let the copper molecules reconnect. Copper wires that break easily when flexed have been improperly annealed or possibly not annealed at all. Repeatedly bending or flexing the conductor, no matter how well it is annealed, can cause any conductor to eventually "work harden" and break.

We depend on the insulation used to cover the wire. As long as the wire is covered and the cable is not damaged, oxidation is rarely a problem.

There are different available grades of copper. The most common high-quality grade is called *electrolytic tough-pitch* (ETP). It can be produced at different purities. Most high-quality cables have a purity of 99.95 percent.

High-purity "oxygen-free" copper

Some high-end audio cable manufacturers advertise "oxygen-free" copper. However, there is no industry standard to define "oxygen free" purity. Purity is often described by the number of "nines." For instance, 99.95 percent purity would be "three nines."

A STORY ABOUT HIGH-END CABLES

A well-known speaker manufacturer, who shall be nameless, once related a story about speaker cables. He assembled a group of high-end aficionados to listen to a number of high-end speaker cables. The group included PhDs, scientists, other manufacturers, even some military brass. The choice of cables was laid in plain sight between a top-of-the-line amplifier and the speakers. The source program material and equipment had been chosen and approved by the group. First, they started with some generic 12-gage "zip cord" as a baseline measurement. Then the helpers would announce the next cable and go behind the amplifier and speaker to change to that cable. Of course, when they came to the super-expensive cables, the difference was dramatic. At the conclusion, although the group disagreed about the details, they all agreed that any of the high-end cables left the 12-gage zip cord in the dust. So the manufacturer who had set up the test had neither the nerve, nor the guts, to tell them that, throughout the entire test, when the helpers went behind the amplifier and speakers, they had changed nothing. The group had been listening to the 12-gage zip cord for the entire time.

The process to make high-purity copper is applied during the refining process. While the copper is in a molten state, a reaction is driven in reverse to purge the copper of sulfur dioxide (SO_2). Only about 0.9 percent of all refined copper is made in this manner. The remaining 99.1 percent of all refined copper has "standard" purity.

Some cables can be four, five, or even six nines. Of course, each of these levels of purity adds significant cost to the final cable. As we will see, there are many parameters that will have a more profound effect on cable performance and electrical measurements.

Can high-purity copper make a difference? The difference in resistance can be measured easily in a laboratory. The difference between three-nines and six-nines copper is about 1.5 percent. It is much easier to see why capacitance will have a greater effect on performance parameters than high-purity copper. See Capacitance on page 50.

Listening to cable

Nevertheless, many high-end manufacturers or assembly houses say that high-purity copper "sounds" better. There are no standards for "listening" to a cable, and because the results of such listening are, at the very least, subjective, this book will concentrate on what can be measured, not what can be heard, in wire and cable.

Of course, this does not prevent anyone from listening to different designs, if they so wish, and forming their own opinions. Unfortunately, there is very little correlation between what is measured in a laboratory and what is heard. One could easily produce a cable with gross variations in parameter, such as very high resistance or high capacitance, and make two cables that could easily be identified in a listening test. But many of the parameters cited in high-end cable literature, such as depth, soundstage, or detail, have no correlation to any known measurable design parameter. Cable manufacturers, despite what many think, would be delighted to make such a correlation. Just imagine the market potential if one could produce a cable with a consistently excellent soundstage or some other factor. Such a correlation between a measured parameter and a heard effect would definitely be a multimillion-dollar breakthrough.

Gold

Gold, of course, is very expensive. And, as can be seen in Table 1-1, its resistance is only average. That being said, there are a number of applications that call for gold wires. One of these is the interconnection inside miniature devices, such as integrated circuits (ICs). Gold can be made into extremely thin wires. Its flexibility and flex-life are unmatched, and it requires no annealing, regardless of the number of drawing steps required. If you want super-tiny wires, and the distance they will go is equally small, the added resistance of gold is a minor factor.

WHEN GOLD IS A GOOD CHOICE

There is one place where gold is a good choice and that is in connectors. Where connector surfaces are joined, especially where they are repeatedly connected and disconnected, gold is an excellent coating choice. Because it does not oxidize, the surface stays clean and the minimal resistance increase of a thin layer of gold is much less than that of some other oxidized metal. Of course, gold is a soft metal so connectors must have sufficient gold layer depth to withstand the wiping of the connector surfaces during connection and disconnection. Many low-quality connectors have a "flashed" surface of 3 to 4 microinches of gold that is easily wiped off with a few insertions. Long-life connectors have 50 microinches of gold, which will last quite a while. Plated gold can be soldered easily. In fact, the solderability of gold is much better than that of many other metals used on connectors, such as nickel. Connectors that are gold plated on the outside, where no contact takes place, are done so purely for appearance. Such exterior plating has no effect on internal performance. Because of its reactivity (Table 1-3) gold ideally should be mated with gold.

Aluminum

For even lower cost than copper, but lower performance, aluminum is the next most common material used. It is inexpensive, easy to work with, and does not require annealing. However, as can be seen in Table 1-1, aluminum has almost twice the resistance of copper. So aluminum is used most often when cost is the major factor.

Low-cost home hi-fi cables often are made with aluminum conductors and aluminum shields. Despite the high resistance, given that they have very short lengths, this is rarely a problem. The foil and braid shields on CATV/broadband cable are most often aluminum to keep the cost and weight down. Because the foil is doing most of the work, using aluminum in the braid is a minor compromise (see CATV/broadband in Chapter 8).

Nickel

Nickel is resistant to oxidation and very hard. For that reason, and because it is a lot cheaper than gold, it is used most often to plate connectors. Although it is not a low-resistance metal, little of it is used on the connector. Nickel has excellent wear characteristics, making it a good choice for connectors that are repeatedly inserted and removed. It is almost never used on wires, because even bare copper can outperform it.

Steel

You might be surprised to learn that, in the design and manufacture of wire and cable, steel is probably the second most common conductor material. Even though it has very high resistance, it has two features that make it popular: it is very easy to make and thus cheap, and, it is very strong.

> ### ONCE UPON A TIME...
> There was a time when almost all the electronic wiring in the world was steel wire. Do you know when this was? Before 1876 and the invention of the telephone. What signals did this steel wire carry? Telegraph. Copper wire was hard to make, expensive, and not very strong. Steel wire was easy to make and very strong. It just had high resistance, so you had to use larger wire or a higher voltage.

You will see steel wires in three types of cable designs. The most common is a steel wire coated with copper. These cables are designed for high frequencies, where only the "skin" of the wire is actually carrying the signal. The steel adds great strength to the conductor without added cost. Copper-clad steel is the standard interior conductor for most cable television/CATV/broadband/satellite coax.

The second place you will see steel wires is in a stranded construction where the steel wires are combined with copper wires. This adds strength and flex-life to the copper wires.

Usually, the steel conductors are copper clad to reduce the corrosion potential between the copper and the steel.

The third application for steel wires is to support other cables, most commonly coax cables, when strung between poles in an aerial application. That steel wire is called a *messenger* and has no other function except to physically support the wire to which it is attached. It carries no signal.

Alloy Conductors

Occasionally, you might see conductors made of alloys that are combinations of metals. In the past, one of the most common was cadmium–bronze. Although it had slightly higher resistance than copper, it had much greater strength and flex-life than copper conductors. Unfortunately, in the processing of the wire, cadmium dust is produced. Cadmium is toxic, so most products no longer use this alloy. There are other metal combinations that approach the strength and flex-life of cadmium bronze, but without the cadmium, such as alloy CT-137.

Sometimes these alloy conductors are silver clad to reduce the resistance, not to protect the wire underneath. In fact, the added silver can take an alloyed wire down in resistance equal to that of a regular copper wire while preserving the strength and flex-life of the alloy conductor. Silver-coated alloys are sometimes used in miniature microphone cables.

Solid and Stranded

Conductors can be made in two basic ways, solid and stranded. Solid conductors are made from one conductor, as shown in cross-section in Figure 1-7. Stranded conductors are made from multiple conductors that are wound, braided, or other-

Figure 1-7 Solid conductor.

wise held together, as shown in cross-section in Figure 1-8. They operate as one conductor. You can get the same gage in a solid or stranded wire. The stranded construction uses multiple bare wires of a smaller gage.

Solid conductors have better electrical performance, especially at high frequencies. Stranded wires are more flexible and have better flex-life.

Figure 1-8 Stranded conductor.

Insulation and Dielectrics

Wires are insulated to prevent them from touching other wires. When wires carry signals, touching other wires can degrade the performance of the cable or short out (short circuit) the signal entirely and prevent any signal from reaching the destination. If there is significant amperage or voltage on the wires, touching wires can be a safety or fire hazard.

When signals are sent down wires, the insulation has an effect on the distance and quality of the signal. Where this effect is of concern, the insulation is called a *dielectric*. The quality of the dielectric is determined in a laboratory test. This test measures the *dielectric constant*. Table 1-4 lists common plastics used as dielectrics and their dielectric constants. The better the plastic, the lower the dielectric constant. In this case, better means less loss, especially at high frequencies. The dielectric constant is a unitless number.

TABLE 1-4   Plastics and Dielectric Constants

Plastic	Symbol	Dielectric Constant
Polyvinyl chloride	PVC	3–8
Polypropylene	PP	2.37
Polyethylene	PE	2.29
Teflon	FEP, TFE, PTFE	2.0
Air		1.0167
Vacuum	(reference)	1

THE STORY OF POLYETHYLENE

Polyethylene was invented by accident in 1933 by E.W. Fawcett and R.O. Gibson of Great Britain. They were experimenting with compressed gasses. When they tried ethylene gas, a by-product of petroleum processing, their container exploded. Inside they found odd white flakes, which turned out to be thermoplastic. Thermoplastic materials can be repeatedly melted and extruded into a shape, such as around a wire. They had squeezed ethylene gas into a long chain of many molecules, or "poly" ethylene. When tested, this new plastic had the best dielectric constant ever seen. So it immediately became top secret and stayed that way throughout World War II. Polyethylene made it possible to make cables that allowed radar to be put into planes. Despite major effort, the Germans couldn't figure out what tree this magic material had come from. No matter, because they simply took the cables out of crashed Allied bombers and, because it was thermoplastic, made their own cables. The world uses more than 100 billion pounds of polyethylene each year. So next time you pull down a bag at the supermarket to put in some veggies, think that there was a time when you could have traded that bag for its weight in gold, and then been arrested!

Table 1-4 also shows air, the best common dielectric, and vacuum, the ideal dielectric, for comparison. Although it is impractical to put a vacuum around a wire, it is very practical to use air. A foam dielectric is a mixture of plastic and air.

The chosen dielectric affects the quality of the cable and will affect many parameters, such as capacitance and impedance. We will examine these and other effects. These are influenced by the frequency of the signal running down the cable.

Shielding

Shielding is used to protect the cable from interference. Shielding is a metal layer wrapped around one or more conductors. Shielding can be used on any number of conductors. It can be used on a single conductor, multiconductors, or twisted pairs. Coaxes are a special form of a shielded single conductor.

Shielding is never perfect. All shields have limitations or inherent flaws based on construction of the shielding elements. Signals can be radiated from pairs or coaxes, or they can pick up external noise or signals from adjacent pairs. This is sometimes referred to as *leakage*. The effectiveness of any shield or combination of shields can be measured and expressed in dBs of rejection. The following section outlines the most common types of shielding and where it is most and least effective. To simplify this discussion, the inside of the cable, whether one or more conductors or a twisted pair, referred to as the "core" of the cable.

Serve or spiral shield

Serve or spiral shields are made of one or more wires simply wound around the core (Figure 1-9). This type of shield is very inexpensive, relatively easy to apply, and extremely flexible. Being a coil of wire, it can have higher inductance than other shields made with wires, so it is used most often

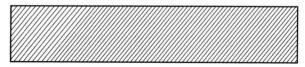

Figure 1-9   Serve or spiral shield.

at low frequencies such as audio. Most low-end consumer hookup cables have spiral shields of aluminum, indicating their low quality. When cables with spiral shields are bent or flexed, the shielding has a tendency to open up, allowing noise through and reducing shield effectiveness. Maximum coverage is 95 percent.

Dual-layer spiral/serve shields

Dual-layer spiral/serve shields are very flexible, and because they cross each other, they short-circuit the inductive problems of a single spiral (Figure 1-10). However, when these cables are flexed or bent, these shields have a tendency to open up, allowing noise in and compromising shield effectiveness. This design is very common on many ultra-flexible snake cables, especially those made in Europe or Japan.

Braid shield

Braid shields are made by interweaving a dual serve of single or groups of wires around the core (Figure 1-11). Because each layer crosses the other, any inductive effect is short-circuited, but beware if the braid is wound so tightly it cannot move when the cable is flexed. This type of shield is effective

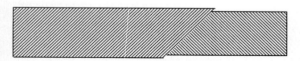

Figure 1-10   Dual spiral/serve or Reussen shield.

Figure 1-11 Braid shield.

well into the megahertz range. It is also an excellent shield for low-frequency protection.

French braid shield

French Braid® shields are proprietary shields made by taking two spiral (serve) layers and interweaving them in one (Figure 1-12). This provides the flexibility of a serve/spiral shield and the mechanical and electrical performance of a braid shield.

Foil shield

Foil is a very common shielding material (Figure 1-13). It is faster to apply than any of the wire type of shields and cost effective. It can be quite flexible, although the wire-style shields can be configured to have greater flex-life.

Figure 1-12 French braid shield.

Figure 1-13 Foil shield.

Unlike braids and serves, foil provides 100 percent coverage. Most foil shields are aluminum foil, although copper foil can be used. Because foil is thin, the resistance difference between aluminum and copper is minimal.

In foil-shielded cables, the low-resistance path is a "ground wire" made of stranded conductors or a braid over the foil. They have the most influence on signal loss or *attenuation*. As long as the shield has a "good" path to ground, shield effectiveness also will be good.

Because foil is thin and easily broken, it is usually glued to a plastic, such as polyester, as shown in Figure 1-13. One side of the foil is a shiny silver-like color. The other side is a colored plastic, blue, green, or red, whatever color is appropriate.

Because foil cannot be crimped or soldered, there is a second conductor physically in contact to make the connection. In coaxes, it is a bare wire braid over the foil. For coax, the foil layer must therefore be turned so it is "foil facing out" to make contact with the braid.

In a multiconductor or twisted pair cable, there is a bare *drain wire* wound in contact with the foil side of the tape, or there is a braid over the foil. It is the braid or drain wire that is used as the connection, not the foil directly. If the foil is applied facing out, the drain wire must be on the outside, as in Figure 1-14. If the foil layer is facing in, the drain wire must be inside, or underneath the foil, as in Figure 1-17.

The cable with the internal drain wire (Figure 1-17) has some surprising advantages over the external drain wire

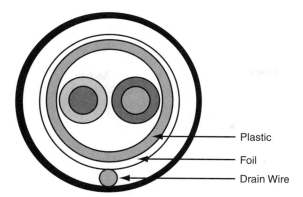

Plastic

Foil

Drain Wire

Figure 1-14   Twisted pair with external drain wire.

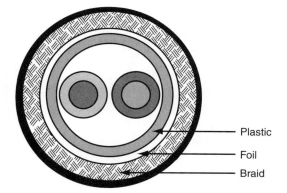

Plastic

Foil

Braid

Figure 1-15   Twisted pair with foil/braid shield.

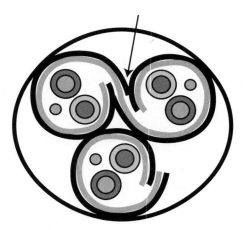

Figure 1-16 Old-style snake cable.

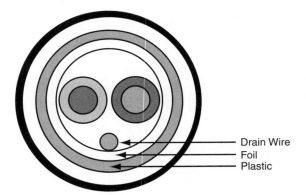

Figure 1-17 Twisted pair with internal drain wire.

(Figure 1-14). First, when the jacket is stripped off (Figure 1-18), one must be careful not to accidentally strip off the drain wire. Because you cannot determine the location of the drain wire before removing the jacket, it is all too easy to nick the drain wire. Nicking the drain wire may cut some of the strands and change its effective gage, increasing resistance and reducing ruggedness. Or you can accidentally strip off the drain wire entirely. Having the drain wire internal, as in Figure 1-17, makes it almost impossible to strip off.

Cable manufacturers have added another feature (Figure 1-18), bonded foil. To understand what bonded foil is and what the advantages are when stripping and preparing cables, we must look at the preparation steps required in Table 1-5.

If the foil were glued, or bonded to the jacket, then removal of the jacket would accomplish all these steps in one motion. This requires that a layer of glue be deposited on the

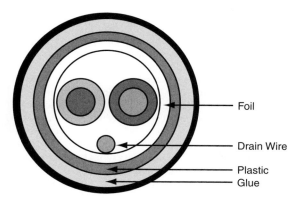

Figure 1-18   Twisted pair with internal drain wire and bonded foil.

plastic side of the foil, that the tape be applied with the foil facing inside, and with the drain wire internal to the cable. The glue is shown in Figure 1-18.

TABLE 1-5   Preparing External Drain Wire Twisted Pair

Step 1	Remove jacket (be careful not to damage the drain wire)
Step 2	Make a nick in the foil where the foil meets the jacket
Step 3	Tear the foil, starting at the nick, and remove it
Step 4	Strip the two wires of the twisted pair, if required
	You now have the twisted pair and drain wire ready

TABLE 1-6   Preparing Internal Drain Wire/Bonded Foil Twisted Pair

Step 1	Remove jacket.
Step 2	Strip each wire of the twisted pair, if required.
	You now have the twisted pair and drain wire ready.

These improvements were especially appreciated by assembly houses that use automatic machinery to prepare cable. But these improvements have also been appreciated by installers and system integrators.

It is not required in any way that a cable with internal drain also have a bonded foil. But you cannot bond the foil of a cable with an external drain or you will glue the drain wire, making it impossible to use. The glue layer is only a few mils thick, so it does not add significantly to the size, weight, stiffness, or "spool-radius memory," a term coined by my friend Joe Metzger at Sony SIC. With identical materials, bonded-foil cables are often the same cost as unbonded.

The bonded-foil internal drain construction saves as much as 30 seconds per cable end. If you have only a few cables to prepare, it's probably not a big deal, but if you are doing a major job with 10,000 pairs, that's 20,000 ends. At 30 seconds per end, that's 10,000 minutes or 167 hours. Multiply that by the hourly rate of an installer and we're talking a serious percentage of the cost of the cable.

If the multiconductor or multipair cable has a combination shield, foil-plus-braid, as shown in Figure 1-15, the preparation for this cable is much more labor intensive. Table 1-7 lists the steps needed to prepare this cable.

TABLE 1-7   Preparing Foil and Braid Multiconductor

Step 1	Remove jacket without damaging the braid
Step 2	Using a pick, unweave the braid
Step 3	Twist the unwoven wires into a single conductor; this woven wire is your ground/shield connection
Step 4	Make a small cut in the foil where it enters the jacket
Step 5	Tear the foil around the cable and remove it
Step 6	Strip the inner wires (if required)
	You are now ready to connect this cable

Shields and multipairs

You can also have multiple shielded pairs inside a single jacket. In those constructions, the foil must always be facing in. If you have two or more pairs with foil facing out and the foils touch, any noise on one foil will instantly be on the foil touching it and will be introduced to the pair under the second foil. You might as well put a foil over both pairs instead of over each pair.

Most pairs with foil shields have the foil turned in, with internal drain wires. But when these cables are flexed or bent, it is still possible for one foil edge to touch another foil edge (Figure 1-16), with the same consequences; that is, increased *crosstalk*.

The arrow in Figure 1-16 shows where the foils are touching. The solution to this problem is to bend the edge of the foil before it is folded. The ideal is called a "Z-fold®." In this design, both edges of the foil are folded and then, as the foil is applied to the pair, the two folds interlock. This ensures that nothing will unfold, no matter how much the cable is bent or flexed.

One common solution to the problem of touching foils is to jacket each pair. Of course, the shields can never come in contact, but the cable becomes bigger, stiffer, and more expensive. Nevertheless, the majority of audio snake cable today consists of individually jacketed shielded pairs. Besides crosstalk protection, individual pairs allow you to split open the cable and send each pair to a different destination, as far as you need, with no reduction in performance.

Splitting open a snake cable of foil-only pairs requires some method of keeping the foil and pairs intact, such as putting heatshrink tubing on each pair. Even if the pairs are separated only by a few feet, this can be a frustrating and laborious undertaking and you'll wonder why you didn't spring for the few dollars extra for the individually jacketed pairs. If all the pairs start in one place and end in one place, especially if they are all in one connector, foil-only pairs might be an appropriate choice.

If size and cost are critical parameters, foil-only pairs can provide significant diameter and price reductions with no reduction in performance. The real problem with foil-only pairs is pair identification. Because the foil consists of a layer of aluminum backed by a layer of polyester, the plastic side can always be colored.

Even so, no manufacturer is going to buy enough different colors of foil to make up a 52-pair audio snake, for instance. Instead, manufacturers will buy three or four colors of foil and then change the inside pair colors. Often it is all one foil color, and it is the pair color you have to keep track of. Each manufacturer of foil-only pairs has a specific

color code you must follow. Keeping track of dozens of different color combinations can require real concentration.

Once you individually jacket the pairs, you can color the
jackets so that you can identify the pairs. The most common
color code for this is the "resistor" color code (Table 1-8).

TABLE 1-8   Resistor Color Code

Color	Number
Black	0
Brown	1
Red	2
Orange	3
Yellow	4
Green	5
Blue	6
Violet (Purple)	7
Gray	8
White	9

Because black indicates 0, it is sometimes not used to color
internal jackets, or it is turned into color 10. Of course, this is
good only for the first nine or ten pairs. After that, most manufacturers use a color with a stripe to indicate the individual
pairs, so you're back to the individual manufacturer's special
color code as it appears on the jacket of each pair.

E- and H-Fields

Electromagnetic fields, whether desired or undesired, consist
of two types: E-fields and H-fields. H-fields are magnetic fields.
Low frequencies such as 60-Hz power and lighting produce H-
fields. It is very difficult to shield against H-fields. Solid steel
conduit is the most common way to protect against H-field

noise. Mu-metal, a special multilayered steel material, is sometimes used to protect against H-field noise. In rare instances, mu-metal can be wound or braided around cable to reduce H-field noise. As we will see, there are other easier, and sometimes more effective, alternatives, such as *balanced lines* or *starquad* cables, both of which are dealt with in Chapter 3.

The most common protection against magnetic H-field noise is distance. Simply move cables away from H-field sources. The inverse-square law applies to H-fields (and to E-fields). When you double the distance from the source of the noise, the noise is reduced to one-fourth of the intensity. Thus, if you have problems with 60-Hz noise getting into audio and video cables because the signal cable is near power cables, simply move the signal cables away from the power cables. If you must cross power cables with audio or video cables, do so at right angles, where the magnetic H-field of the power cable aligns as little as possible with the signal cable fields.

In contrast, E-fields generally produce high-frequency effects. Braid shields are good at shielding against E-fields and remain effective up to hundreds of megahertz. Foil shielding starts to be effective at 10 MHz and becomes the most effective of all at higher frequencies.

EMI and RFI

H-field interference and noise are sometimes referred to as electromagnetic interference or EMI. E-field interference and noise are sometimes referred to as radio frequency interference or RFI.

Shields and Frequency

Each shield type has good and bad points. Most perform-ance differences have to do with their effectiveness at different frequencies. No common shield has any effect below 100-Hz for H-fields, which explains why signal cables are particularly susceptible to low-frequency noise, such as 60-Hz power and lighting cables. The only protection against low-frequency H-fields are conduits, balanced lines, or star-quad configurations. As mentioned previously, you also can move the cable away from the noise source. Shields still work adequately with the decreasing E-fields at those lower frequencies.

Conduit

Solid steel conduit is one of the few effective mitigations against 60-Hz H-field interference, which will be discussed in much greater detail in the Installation section (Chapter 8). Solid steel conduit can be effective at 60-Hz up to 27 dB, assuming good installation practices, especially good conduit joints.

Little or no effective shielding can be expected from flex conduit, such as interlocking steel conduit, or any non-metallic conduit, such as PVC or innerduct linings.

Balanced lines

A good balanced line may be effective against 60 Hz. The amount of 60-Hz rejection will be based on the *common-mode rejection ratio* (CMRR) of the source and destination devices and the balance of the twisted pair. A detailed engineering analysis would be required to determine the actual benefit, so you could hire an engineering consultant to advise you. Or you could buy or rent the appropriate test equipment. Perhaps the most practical way to determine effectiveness is by trial and error.

Starquad

Four-conductor starquad cables, most commonly used for microphone wiring, are dramatically effective at 60 Hz. Ultimately, starquad effectiveness is based on the CMRR of the source and destination devices, just like standard balanced lines. However, the high capacitance of starquad limits the distance these cables can go and may be self-defeating for installations with longer cables. See more on starquad cable on page 110.

Above 1000 Hz

Braid shield effectiveness is best against E-field noise. For that reason braid shields begin to be effective at about 1 kHz. At these low frequencies, braids are the most effective choice, based largely on the resistance or gage of the braid. They provide a low-resistance path to ground for noise that is intercepted.

Braids, by the very nature of their design and construction, provide less than 100 percent coverage. The best braid provides approximately 95 percent coverage. Even two layers of high-coverage braid provides only 98 percent coverage.

Moreover, as frequencies go higher, those natural "holes" in the braid appear to get electrically larger and larger. As the wavelengths of the increasing frequencies become smaller and smaller, the openings in the braid become a more significant percentage of a wavelength. It is there that a foil shield becomes superior. This transition between braid and foil effectiveness begins at around 10 MHz.

Foil shields lack the large gage and low resistance of a braid shield. Foils are, in reality, a skin-effect shield. Foil, with a depth of 2 to 3 mils, only begins to be effective in the megahertz range and doesn't become the major noise-reduction mechanism until 10 MHz. Foil does not reach full effectiveness until 50 MHz or so, when the skin effect becomes more important than low resistance.

Cables assembled with foil shields can help reduce *capacitive coupling* at low frequencies and high frequency (RF) shielding. Twisted-pair cables, which often feature foil shields, depend on the twisting of the pairs and the CMRR of the source and destination devices for a significant part of the noise rejection.

Multiple-Layer Shielding

For the ultimate in shielding, combinations of braid and foil are commonly used. Many cables have a braid shield (95 percent coverage) over a foil shield (100 percent coverage).

As we will see, CATV/broadband cables often have multiple layers of shielding amounting to a "quad" shield, which consists of foil, braid, foil, braid. Because television channels do not start until channel 2 (54 MHz), there is no need for low-frequency coverage, and a foil shield is an excellent choice. The braid around it simply provides the connector something to grab onto and a low-resistance path for the noise.

Shield Effectiveness

Shield effectiveness is a specific measurement made in the laboratory. It is most often made on CATV/broadband cables because of signal emission limits imposed by the Federal Communication Commission (FCC). The FCC has established the maximum signal that can be "leaked," or radiated, by a cable, so cables must be tested to meet those guidelines.

Table 1-9 shows the shield effectiveness for various combinations of foil and braid commonly used for CATV/broadband coaxes. Effectiveness is shown in two frequency bands. Cables shown are typical RG-6 constructions for this application.

TABLE 1-9   Broadband Shield Effectiveness

		Shield Effectiveness	
Shield Type	Construction	5–50 MHz	50 MHz–GHz
Foil/braid	100 percent foil/ 90 percent braid	75 dB	85 dB
Foil/braid/foil	100 percent foil/ 80 percent braid/ 100 percent foil	105 dB	90 dB
Foil/braid/foil/braid	100 percent foil/ 60 percent braid/ 100 percent foil/ 40 percent braid	105 dB	110 dB

Jackets

Jackets are extruded over shielded or unshielded cables. The intent of the jacket is to keep the core elements together and protect them from the rigors of installation or the roughness of use. Outdoor or direct-burial cables require special jacket constructions.

Most often, a jacket extruded over the core and shield has no effect on the electronic performance of a cable. There are a couple of rare exceptions. When the chemical constituents of the plastics that make up the jacket are of such low quality that they begin to separate, they can move, or "migrate," into the shield and sometimes even into the core. This can hasten oxidation or corrosion of the braid.

If the chemical constituents migrate into the core, especially in a coax, it can change the dielectric constant, which would change the impedance of the cable increasing the return loss, and possibly rendering the cable inoperative. This is called *compound migration* and most often occurs in low-cost cables. It sometimes can be detected by a color change in the dielectric from white to yellow.

Table 1-10 lists common jacket materials. As in other applications, each has its good and bad points, which are summarized in the table.

TABLE 1-10   Jacket Compounds

Insulation	Jacket Qualities
Polyvinyl chloride[1]	Inexpensive, flexible, easily colored
Polyethylene[1]	Rugged, outdoor only, poor fire resistance
Teflon®[1]	Rugged, expensive, excellent fire resistance
Rubber[2]	Rugged, moderately expensive, poor colors
Silicon[2]	Rugged, excellent temperature extremes
Hypalon®[2]	Rugged, moderately expensive
Neoprene®[2]	Rugged, moderately expensive
EPDM[2]	Rugged

[1]Thermoplastic type.
[2]Thermoset type.
EPDM, ethylene-propylene-diene monomer.

There are two types of jacket materials, thermoplastic, and thermoset. Thermoplastic materials can be extruded and cooled. Later they can be ground up, heated, and extruded again. This is the original recycling and the reason they are called *plastic*. Thermoset materials are cured, usually by heat or steam. When finished, they cannot be recycled into new products, which is the reason the world has such a problem with old tires. They cannot be easily or effectively recycled.

Jacket colors

Jacket colors can be a significant help in installation. Colors can be used to identify a single cable, groups of cables, or even whole portions of an installation. Most cables are made with PVC jackets. PVC is easy to color and can be made with solid, vibrant colors.

There are only a few standard color codes for audio or video applications. Red cables are often used for specific alarm-security applications. These are rarely part of an audio/video installation, so it is unlikely that you will encounter them. For RGB analog video monitors or VGA, S-VGA, or X-VGA computer monitors, red, green, and blue colored jackets are often used on the internal coaxes to aid in connection of these cables. In Europe, purple (violet) often indicates digital signals, especially digital audio. In home audio/video, yellow indicates video, and red and white indicate the left and right audio channels, respectively.

Aside from these colors, there is no standard, which means you can create your own. If you are part of a group, station group, or consortium, you might want to create a color-code standard for your group. That way, engineers from other facilities will not be confused if they have to move between installations.

Although there are millions of possible colors which can be mixed or matched, the most common colors available start with the resistor color code shown in Table 1-8. You can use these colors to indicate anything you wish. For instance, a red cable may mean it is from studio 2 (red = 2) or the second floor of a building. Or red could mean "on air—don't unplug." Or red could mean, "This is the boss's computer. Unplug it and you're fired!"

Resistance

Resistance is one type of opposition to the flow of electricity on a conductor. Resistance converts electrical flow into heat. The higher the resistance, the greater the conversion into heat. All other things being equal, the smaller the wire, the greater the resistance. Therefore, resistance is directly related to the size, or gage, of a wire. The unit of resistance is the ohm (Ω).

GRANDMA'S CHRISTMAS TREE

Remember Grandma's Christmas tree? It had a thousand lights on it, and all of them fed off one 18-gage extension cord. Pick up that extension cord and what will you find? Mostly likely, the plastic will be dripping off the cord and into your hand. It will not be a pleasant experience. We assume that the two wires had not already shorted together and blown a fuse or circuit breaker. This also shows one factor about insulated cable: the amperage limits, and therefore the operational temperature limits, is set by the insulation, not by the wire.

Table 1-11 lists wire sizes, called the American Wire Gage (AWG), and the resistances of those wires. A more complete chart can be found in the Addendum. Solid and stranded wires of the same gage have different resistances because the solid wire is completely copper, whereas stranded wire has spaces (interstices) between the strands. Resistance changes with temperature. The resistance values shown in Table 1-11 are for conductors at 68°F.

TABLE 1-11   Wire Gage and Resistance

| | Resistance Ω/1,000-ft | |
AWG	Solid wire	Stranded wire
30	103.2	112
28	64.9	70.7
26	40.81	44.4
24	25.67	27.7
22	16.14	17.5
20	10.15	10.9
18	6.385	6.92
16	4.016	4.35
14	2.525	2.73
12	1.588	1.71
10	0.9989	1.08

Where signals of only a few milliamps are traveling down a wire, the heat generated by resistance is imperceptible. It is only when you have high amperage that you can actually "feel" the heat generated, and only when the wire is small enough to have significant resistance.

Heat generated at a specific place in a cable, such as in very large coaxes that connect transmitters to broadcast antennas, indicates a place of high resistance. This heat might indicate a poorly installed connector, a flaw in the cable construction, or some other cause of high resistance. When not expected, heat is a danger sign.

Frequency and Bandwidth

As the word suggests, frequency is a measure of how frequently something happens. One form of frequency is called a *sine wave* and can be graphed as shown in Figure 1-19.

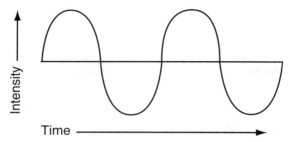

Figure 1-19   Sine wave.

Figure 1-19 shows the flow of electricity on a wire, first in one direction and then reversing and flowing in the other direction. This is also known as *alternating current*, or AC. Figure 1-19 shows two full cycles. If this occurs in 1 second, this is a frequency of two cycles per second, or 2 Hz.

Electricity that comes out of the AC jack on the wall has a frequency of 60 cycles per second, or 60 Hz. That means it reverses direction 60 times every second. An FM radio station might broadcast at a frequency of 100.1 MHz, or 100.1 million times per second. That means, on a radio antenna, a tiny electromagnetic field is being induced, going back and forth 100 million times each second.

Often there are combinations of frequencies. For instance, that FM station at 100.1 MHz carries one or two audio channels (music and voice) of low frequencies and might include "subcarrier" data at intermediate frequencies.

So let's start with a quick analysis of the spectrum shown in Table 1-12—the entire band of frequencies that can affect audio, video, and other related applications.

To understand Table 1-12, be aware that there are hertz (Hz), thousands of hertz or kilohertz (kHz), millions of hertz

TABLE 1-12 Frequencies and Bandwidths

Bandwidth	Application	Considerations	Cable Type
DC Direct Current	Power distribution, basic control circuits	Distance determined by resistance of wire; see Ohm's Law in the Addendum	Any cable, but most often single or multiconductor
20 Hz–20 kHz	Analog audio frequencies	Range of human hearing; some argue that the range should extend below and above these frequencies, such as 10 Hz–50 kHz	Coaxial cable for consumer applications; twisted pairs for professional installations
DC–4.2 MHz	Analog baseband video, such as the video output of a VCR	High frequency requires precision cable	Coaxial cable
DC–24.576 MHz	AES and S/PDIF digital audio	Different sampling rates require different ranges (bandwidths); 24.576 MHz is for 192-kHz sampling, the highest in the AES standards	Standards exist for twisted-pair or coaxial cable

DC–135MHz	SDI (serial digital interface) professional digital video; this bandwidth is for a 270-Mbps (megabits-per second) component, serial digital	Cable and passive components should be tested to the third harmonic of 135 MHz or ~400 MHz	Coaxial cable
DC–750 MHz	Professional uncompressed high definition video (HDTV)	Cable and passive components should be tested to the third harmonic of 750 MHz or 2.25 GHz	Coaxial cable
50 MHz–1 GHz	Standard CATV/ broadband delivery (158 cable channels)	Cable and passive components should be tested to 1 GHz	Coaxial cable

or megahertz (MHz), and billions of Hertz or Gigahertz (GHz). The range of frequencies in any application is called the *bandwidth*.

The range of frequencies covered by audio and video applications can extend into the gigahertz range. Cables designed for operation at those high frequencies can be very different from those intended for low-frequency applications.

Wavelength

As shown in Figure 1-19, alternating signals look like waves. In fact, they are waves especially when they are radiate from an antenna, when they are electromagnetic waves. Each frequency has a corresponding wave, and each wave for a specific frequency has a measured electrical length, or wavelength, as shown in Figure 1-20. The arrow shows one complete wave.

We can determine the wavelength of any given frequency with a simple formula (Figure 1-21), where Wm is the wavelength in meters and F is the frequency in hertz. To convert the answer to feet, just multiply the meters by 3.28.

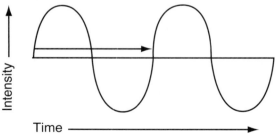

Figure 1-20 Wavelength.

$$W_m = \frac{300,000,000}{F}$$

Figure 1-21   Formula: wavelength.

The *critical distance* is one quarter wavelength. Table 1-13 shows the wavelength and quarter wavelength distances at various frequencies. The distance shown is calculated as if the wire were in free space, or a vacuum. The distances range from thousands of miles to mere inches. In the same way, signals, and the cables and other passive parts through which they travel, become more and more critical as wavelengths get shorter and as frequencies increase.

TABLE 1-13   Frequency and Wavelength

Frequency	Wavelength	Quarter Wavelength
20 Hz	9,318 miles	2,330 miles
60 Hz	3,106 miles	777 miles
20 kHz	9.32 miles	2.33 miles
1 MHz	984 ft	246 ft
4.2 MHz	234 ft	59 ft
6 MHz	164 ft	41 ft
25 MHz	39 ft, 4.32 in	9 ft, 10 in
50 MHz	19 ft, 8.16 in	4 ft, 11 in
100 MHz	9 ft, 10 in	2 ft, 5.52 in
135 MHz	7 ft, 3.47 in	1 ft, 9.87 in
750 MHz	1 ft, 3.74 in	3.94 in
1 GHz	11.8 in	2.95 in
2.25 GHz	5.28 in	1.31 in
3 GHz	3.94 in	0.984 in

> **ACOUSTICAL WAVELENGTH**
>
> More than a few music professionals dispute the extraordinarily long wavelengths for low frequencies because they're thinking of *acoustical wavelength*. A vibrating string or organ pipe is made to produce a specific acoustic frequency and wavelength. That formula is:
>
> $$W_{ft} = \frac{1127}{F}$$
>
> where W_{ft} is the acoustical wavelength in feet, and F is the frequency in hertz.

Wavelength is important because, as frequencies rise, the length of the cable can be less than one wavelength. The general rule of thumb for wavelength is that one-quarter of a wavelength is the critical distance. Quarter wavelength distances are shown in Table 1-3. How likely is it that you will deal with cables that are a quarter of a wavelength or more in length? Unlikely until you reach the megahertz range and very likely above 100 MHz. These distances will be especially important when we discuss impedance.

Velocity of Propagation

Many readers may be surprised to learn that signals traveling down cables do not travel at the speed of light. In fact, the speed of the signal, called the velocity of propagation (V_p), is much less than the speed of light. V_p is related to the dielectric around the conductors and its dielectric constant. Most manufacturers list the V_p for cables designed for high frequencies. If you need to know the dielectric constant instead, you can easily convert by using the formulas shown in Figures 1-22 and 1-23.

How Fast Is Fast?

Only one thing moves at the speed of light, and that is light itself. And it only moves at that speed in one place, in a vacuum, such as outer space. Once light hits our atmosphere, air slows light down to 99.18 percent (the dielectric constant of air = 1.0167). This is so close to 100 percent, or a dielectric constant of 1, that air is usually considered to have a V_p of 100 percent. Most people mistakenly assume that fiber optics, because those signals are light and not electromagnetic, move at the speed of light. In fact, fiber optic cables are even worse than most copper cables, with a velocity of 50 percent or even less. However, fiber offers immense bandwidth and distance, regardless of the "speed" of the signal.

Figure 1-22 shows how to convert the dielectric constant (DC) to the V_p, and Figure 1-23 shows how to convert V_p to the dielectric constant.

Table 1-14 is the same as Table 1-4, except that it includes the equivalent V_p. These are all solid plastics,

$$V_p = \frac{100}{\sqrt{DC}}$$

Figure 1-22   Formula: dielectric constant to velocity of propagation.

$$DC = \frac{10,000}{V_p{}^2}$$

Figure 1-23   Formula: velocity of propagation of dielectric constant.

except for air and a vacuum. When air is added to plastic, making foamed plastics, the dielectric constants can be improved.

Remember how we could compare low-frequency cables by looking at the capacitance? High-frequency cables can be compared by looking at the V_p. You can compare many cables, such as video cables or RF cables. If all other things are equal, V_p is the deciding factor regarding electrical performance.

TABLE 1-14   Plastics, Dielectric Constant, and Velocity

Plastic	Symbol	Dielectric Constant	Velocity of Propagation
Polyvinyl chloride	PVC	3–8	58–35 percent
Polypropylene	PP	2.37	65 percent
Polyethylene	PE	2.29	66 percent
Teflon	FEP, TFE, PTFE	2	70 percent
Air		1.0167	99.7 percent
Vacuum		1	100 percent

Delay

As we know from V_p there is a "delay" for the signal to move from one end of the cable to the other. Figure 1-24 shows the formula for delay. D_n is the delay in nanoseconds (billionths of a second). It is directly related to the dielectric constant and therefore to the V_p.

$$D_n = \sqrt{DC} = \frac{100}{V_p}$$

Figure 1-24   Formula: dielectric constant or velocity to delay.

Signal delay is not a significant factor until you have reached high frequencies or your cable is carrying a digital signal. Then it can be important that signals arrive at the same time. In those applications, you can obtain the delay numbers from the manufacturer or convert the V_p or dielectric constant (Figure 1-24).

Basic delay, based on dielectric constant, should not be confused with the skew/delay often mentioned with premise/data (computer) cables or with the time delay, also called *timing*, used with multicoax RGB or VGA cables.

Wavelength and Velocity of Propagation

The actual wavelength of a signal in a cable, also known as the electrical wavelength, is also determined by the V_p. Multiply the wavelength determined by the equation in Figure 1-21 by the V_p. Table 1-15 shows the distance based on the same frequencies as those shown in Table 1-14, but with a number of common velocities.

Skin Effect

As frequencies get higher, the signal begins to move to the outside of the conductor. At 50 MHz, most of the signal is on the skin of the conductor and the resistance of the wire becomes less important. Figure 1-25 shows a formula for

$$D_{in} = \frac{2.61}{\sqrt{F_{HZ}}}$$

Figure 1-25 Formula: skin depth.

TABLE 1-15   Frequency and Velocity of Propagation

Frequency	Quarter Wave	V_p 83 percent	78 percent	66 percent	50 percent
20 Hz	2,330 miles	1934 miles	1817 miles	1538 miles	1165 miles
20 kHz	2.33 miles	1.93 miles	1.82 miles	1.53 miles	1.17 miles
1 MHz	246 ft	204 ft	192 ft	162 ft	123 ft
4.2 MHz	58.57 ft	48.6 ft	45.68 ft	38.66 ft	29.29 ft
6 MHz	41 ft	34 ft	32 ft	27 ft	20.5 ft
25 MHz	9.84 ft	8.16 ft	7.68 ft	6.49 ft	4.92 ft
50 MHz	4.92 ft	4.08 ft	3.84 ft	3.25 ft	2.46 ft
100 MHz	2.46 ft	2.04 ft	1.92 ft	1.62 ft	1.23 ft
135 MHz	1.82 ft	1.51 ft	1.42 ft	1.2 ft	1.41 ft
750 MHz	3.94 in	3.27 in	3.07 in	2.6 in	1.97 in
1 GHz	2.95 in	2.45 in	2.3 in	1.95 in	1.48 in
2.25 GHz	1.31 in	1.09 in	1.02 in	0.86 in	0.655 in

approximate skin effect on copper conductors, where Din is skin depth in inches and F is frequency in hertz.

Although resistance is less important, the size of the conductor is critical. As the gage of the conductor increases, the surface area, or skin, of the conductor also increases. Therefore, bigger wire can go farther with less loss than smaller wire, even at very high frequencies.

Table 1-16 compares the skin depths at various frequencies to the diameters of different gages of wire and shows what percentages of the conductor are used at those frequencies.

Note that there is a perceptible change from a whole conductor at the upper end of the audio spectrum, but that the significant skin effect, less than 5 percent of the conductor, doesn't begin until 100 MHz for a 24-AWG conductor. For analog audio, this means that all-copper conductors should be used. Copper-clad steel would be an especially bad choice because most audio frequencies use the entire conductor.

This data in Table 1-16 is based on solid wire because that is more commonly used at high frequencies. Stranded is always larger than solid, so the percentages for stranded would be smaller than the percentages shown in Table 1-16.

Capacitance

When two pieces of metal are separated by a nonconductor, an electrical device called a *capacitor* is formed. A capacitor stores voltage. The effect is called *capacitance* and is measured in farads. In the case of two metal wires separated by plastic, the capacitance is measured in picofarads (pF), or trillionths of a farad.

How can such a small amount of anything have an effect? The reason is that the capacitance increases along the length of the cable. For instance, a 1,000-ft cable with 30 pF/ft will have a total capacitance of 30,000 pF, a significant amount.

TABLE 1-16 Frequency, Gage, and Skin Depth

	26 AWG	24 AWG	22 AWG	20 AWG	18 AWG	16 AWG	14 AWG	12 AWG	10 AWG
1 kHz	100%	100%	100%	100%	100%	100%	100%	100%	100%
5 kHz	100%	100%	100%	100%	100%	100%	100%	91%	73%
10 kHz	100%	100%	100%	100%	100%	100%	81%	65%	44%
20 kHz	100%	100%	100%	100%	92%	73%	58%	46%	36%
50 kHz	100%	100%	92%	73%	58%	46%	37%	29%	23%
100 kHz	100%	82%	65%	52%	41%	33%	26%	20%	16%
500 kHz	46%	37%	29%	23%	18%	15%	12%	9%	7%
1 MHz	32%	26%	21%	16%	13%	10%	8%	7%	5%
5 MHz	23%	18%	15%	12%	9%	7%	6%	5%	4%
10 MHz	15%	12%	9%	7%	5%	5%	4%	3%	2%
50 MHz	10%	8%	7%	5%	4%	3%	3%	2%	2%
100 MHz	5%	4%	3%	2%	2%	1%	1%	0.9%	0.7%
1 GHz	3%	3%	2%	2%	1%	1%	0.8%	0.6%	0.5%

Capacitance has one significant bad point. Even though the capacitance is always the same on a given length of cable, it reacts to the frequency on the wire; this effect is called *capacitive reactance*. This effect opposes current flow, so it is also measured in ohms (Ω), like resistance. Resistance affects all frequencies equally, whereas capacitance affects high frequencies more than low frequencies.

By definition, all cables have two or more conductors, so all cables have capacitance. One can substitute a different plastic with a lower dielectric constant and get lower capacitance or move conductors farther apart to reduce capacitance. Every cable will have a "curve," sometimes called the *frequency response* as shown in Figure 1-26 caused by the capacitance.

The amount of acceptable capacitance depends on the frequencies running on the cable and the length of the cable. As we discuss each kind of signal, we will also talk about appropriate capacitance.

Inductance

There is one other effect, called *inductance*, that occurs in a wire. When electricity passes down a wire, an electromag-

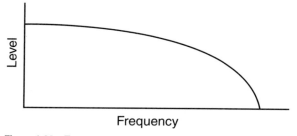

Figure 1-26   Frequency response.

netic field is generated around the wire. Because the wire itself is not magnetic, the field collapses as soon as the electrical potential is removed. If the signal on the wire is alternating, the field builds and collapses at the same rate as the frequency running on the cable. This means that the wire will store some of the energy in its magnetic field. This "field storage" is called "inductance" because the field is "induced" by the electrical flow down the wire. Inductive fields from one cable can interfere with those of other cables nearby.

In multipair cables, signals on each pair can interfere with those on other pairs within the same cable. This effect is called *crosstalk* and can pass inductively, by generating a magnetic field, or capacitively, by forming a capacitor between two conductors within the two cables.

Unlike capacitance, inductance on cable typically is so small that it often is not even listed in wire and cable catalogs. In fact, to get any significant inductance, you have to tightly coil a wire. So a straight piece of cable produces only a very tiny amount of inductance. Similar to capacitance, inductance has an effect based on frequency called *inductive reactance*.

Inductive reactance is different in that it is affected more by low frequencies than high frequencies, the reverse of capacitive reactance. In fact, these two reactances can cancel each other out. Nevertheless, capacitive reactance is more prominent, has a much greater effect, and overwhelms the inductive reactance.

Inductive reactance also opposes electrical flow, so it is also measured in ohms just like resistance and capacitive reactance. Where capacitance is affected by the choice of dielectric, inductance is affected by the size of the wire. The larger the wire, the greater the inductance. This is the reason speaker cables, especially high-end speaker cables, sometimes consider the inductance in addition to other factors. When speaker cables are very large, for example, larger than 10 AWG, inductance should be considered. Also, capacitance on paired cables with large conductors is usually very low, often less than 20 pF/ft. This makes inductance more prominent. See the Addendum for the formulas to measure capacitive reactance and inductive reactance.

Impedance

Impedance, as the word implies, describes an effect that impedes the signal flow. Impedance is the total effect of resistance, capacitance, and inductance. Because each of these oppose electrical flow on a cable in a different way, impedance is a number that describes the combined effect or the total opposition to current flow.

It is also important to understand that resistance is a real effect. It turns electrical flow into heat. Reactance is imaginary. Neither inductive reactance nor capacitive reactance actually "use up" any flow. They merely store it and then release. Although the effect is "imaginary, the result on a signal going down the cable is very real. This is the reason the formulas for impedance contain imaginary numbers to describe the reactance.

Every cable has impedance, but many cables do not have an impedance measurement listed by the manufacturer. There are two reasons for this. First, impedance is not important until the signal on the cable has a high frequency, at least 1 MHz. Therefore, analog audio cable manufacturers do not list impedance values for those lower frequency cables. Impedance is important only when the signal is a quarter of a wavelength at the frequency of operation. Table 1-13 shows how long quarter wavelengths are at different frequencies. Second, the impedance changes at low frequency.

This is not to say that the impedance at low frequencies is of no consequence, but the cable must be long enough to approach a quarter-wavelength. Power-generating companies are very interested in the impedance of cables at 60 Hz. Their transmission lines can easily approach a wavelength at 60 Hz (3106 miles). This is the reason there has been so much work on super-high voltage direct current lines. These avoid the reactive losses at 60 Hz.

Figures 1-30, 1-31, and 1-32 show three formulas for determining the impedance and the mathematical effect of low and high frequencies. The first formula applies where resistance is a factor. As frequencies increase, skin effect takes over, and the resistance becomes less and less a factor. The impedance settles to a value called the *characteristic impedance*, which is the last formula shown. There is a transition between these two values, with the formula for those transition values.

Occasionally, you might see an analog audio cable with an impedance listed in a manufacturer's catalog. Figure 1-27 shows that impedance at low frequencies changes, getting higher as the frequencies decrease. Therefore, the listed impedance must be the characteristic impedance, which

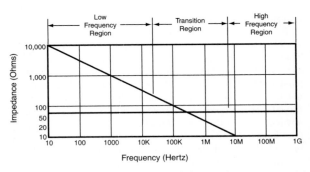

Figure 1-27 Impedance at low and high frequencies.

does not change. This impedance value is listed only for those who intend to use the is cable in the megahertz region and above, not for analog audio.

You can see why impedance at low frequencies doesn't matter. It starts at infinity at DC (a frequency of 0 Hz) and decreases in value until it reaches the characteristic impedance. In the example, this occurs near 10 MHz. If you want a speaker cable with an 8 Ω impedance, you need to ask, "At what frequency should it to be 8 Ω?"

Of course, at any other frequency, such a cable would have a different impedance, but that's fine, because it doesn't matter what the impedance of the cable is at analog audio frequencies. As shown in Table 1-13, any analog audio cable, including speaker cable, would have to be at least a mile long before the impedance becomes important. So what is the impedance of a speaker cable? It doesn't matter what it is until it is long enough to make a difference.

Impedance can be determined by the dimensions of the cable and choice of dielectric. This is shown in Figure 1-28, a cross-section of a coax and Figure 1-29, a cross-section of a twisted-pair cable.

If one knows the size of the center conductor, the distance from the center conductor to the shield, and the dielectric constant of the material in between, one can also determine the impedance of a coax. Note that the size of the center conductor determines the resistance and the inductance. The distance between conductors, with the material in between, determines the capacitance. The actual formula is in the addendum.

The twisted-pair formula is virtually identical. Three measurements determine impedance. First is the size of the conductor. We are assuming that the conductors have the same gage. Second is the distance between conductors. The third is the dielectric constant of the material in between. The actual formula is in the addendum.

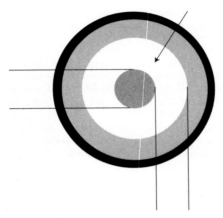

Figure 1-28 Impedance of a coax.

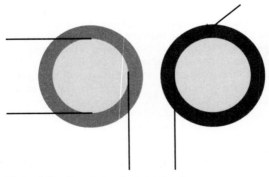

Figure 1-29 Impedance of a twisted pair.

For coaxes, the distance between conductors is "locked in" by the construction. With the twisted pairs, the distance between conductors can change when the cable is flexed, so the impedance is not as stable in a twisted pair as it is in a coax.

There are cutting-edge twisted pairs where the conductors are bonded together. This is a proprietary process where the conductors are extruded and joined together at the same time. These bonded twisted pairs can approach coaxes in impedance stability, but coax is still superior. Bonded twisted pairs, although originally intended for premise/data computer applications, also are being used for non-data applications such as audio and video. We will discuss each of these in the following chapters.

There are three simplified formulas for impedance. They are shown in Figures 1-30, 1-31, and 1-32. Figure 1-30 shows the formula for low frequencies, where resistance is a factor. As the frequency increases and skin effect becomes a factor, resistance is less and less of a consideration. Eventually only the capacitance and inductance are serious enough to consider and the formulas in Figure 1-31 become important. The transition between these two formulas is shown in Figure 1-32.

$$Z_0 = \sqrt{\frac{R}{j2\pi fC}}$$

Figure 1-30 Formula: transition impedance between low and high frequencies.

$$Z_0 = \sqrt{\frac{L}{C}}$$

Figure 1-31 Formula: impedance at low frequencies.

$$Z_0 = \sqrt{\frac{R + j2\pi fL}{G + j2\pi fC}}$$

Figure 1-32   Formula: impedance at high frequencies.

Figure 1-31 shows the simplest formula, the characteristic impedance, the value that applies for all high-frequency applications. In these formulas, Z_0 is the impedance in (Ω), R is the resistance (Ω), j indicates the complex (imaginary) number, pi (π) is 3.14159, f is the frequency (Hz), C is the capacitance (farads), L is the inductance (henrys), and G is the conductance per unit length (mhos).

Foam Dielectrics

You can change the performance of a cable by using a dielectric with a better dielectric constant, or you can mix plastics with the best dielectric available, which is air. Of course, vacuum is the best dielectric of all (dielectric constant = 1), but it is hard to mix a vacuum with a plastic. However, air has a dielectric constant of 1.0167, which is pretty close to that of a vacuum. Mixing air with plastic is much easier and produces foam.

There are two ways to foam plastic. The first is chemical foam, which is accomplished by mixing pellets of foaming agent with pellets of plastic. These are precisely mixed, heated, melted, and extruded onto a wire. The heating process starts the foaming process. This will take solid polyethylene from a V_p of 66 percent to a V_p of 78 percent. When the chemical foaming process is done, the dielectric contains plastic, air, and all the chemicals left from the reaction. These residual parts negatively affect the dielectric constant and velocity of the foam.

What would be ideal is a foam that consists of plastic and air and nothing else. We could be even more precise and say plastic and nitrogen. Air is mostly nitrogen. Using nitrogen would avoid particulates in the air, resist moisture, and have absolute control over the resulting mixture. This technique produces *nitrogen gas-injected foam*.

Nitrogen gas-injected foam originally was invented to decrease loss and improve high-frequency response for CATV/broadband cables. Because they operate in the gigahertz band (very high frequencies), this was a real improvement in performance. There is only one problem with foam—it is soft.

High-Density Hard-Cell Foam

Air and plastic mixed into foam is soft. Soft foam is easily deformed and allows the center conductor to move when the cable is bent or flexed. This movement affects the distance between the center conductor and shield, which is one of the three critical factors that affect impedance (Figure 1-28).

Changes in impedance affect how signals, especially high-frequency signals, pass down the cable. Impedance variations cause high-frequency signals to reflect back to the source. This is called *return loss*.

To prevent return loss, the foam should be hard. Figures 1-33 to 1-35 show how to harden foam. Figure 1-33 shows what a common foam might look like. Notice the variations in bubble size. It would be impossible to predict what the dielectric constant of such a foam would be, what percentage is air, and what percentage is plastic.

What you need is a consistent foam, as shown in Figure 1-34. This foam is predictable in terms of plastic content and air (or nitrogen) content. The problem with this foam is that it is very soft. You can check this with foam cables, especially foam core coaxes. Just press the foam dielectric between

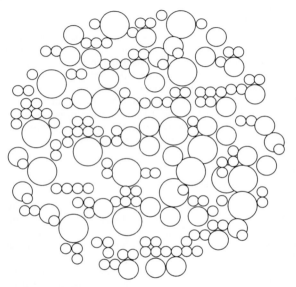

Figure 1-33   Generic foam.

your fingers. If you can deform the cable with just your fingers, imagine what a bend or flex will do, what would happen if someone stepped on the cable, or put a heavy weight on the cable.

The ideal foam is shown in Figure 1-35. This is made of many uniform small cells. Harder, high-density plastic, such as high-density polyethylene, is combined with nitrogen gas to make bubbles where the surface area of the bubbles is much greater than that of standard foam (Figure 1-33). This is high-density hard-cell foam, the cutting edge of foaming technology.

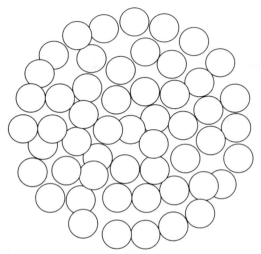

Figure 1-34 Consistent foam.

Recent work with high-density hard-cell foam has produced consistent foam that exceeds 86 percent V_p and yet
maintains excellent crush resistance. A higher V_p foam means
that a cable will have less loss at high frequencies than one
with lower V_p. Alternatively, you could choose to make a
smaller cable with the same properties as a previous version.

Many manufacturers make foam cores, and the most
common problem is consistency. Very few make the cores the
same way the next day, the next week, or the next year.
Consistency is the key derived feature and results in predictable cable performance (Table 1-17). Inconsistent foam
can result in a wide range of cable performance problems,
such as variations in capacitance, variations in impedance,

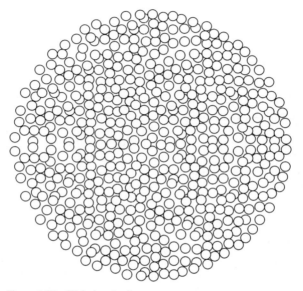

Figure 1-35   High-density foam.

TABLE 1-17   Velocity of Propagation and Dielectric Constant

Velocity of Propagation	Dielectric Constant
82 percent	1.48
83 percent	1.45
84 percent	1.42
85 percent	1.38
86 percent	1.35

and increased return loss, all of which we will discuss in Chapter 7.

Return Loss

When cable lengths are longer than one-quarter wavelength, the impedance of the cable becomes important. Variations in impedance cause the signal to reflect back or return to the source device.

Return loss occurs when the signal is returned to the source and causes loss of signal at the destination. It is measured in decibels (dB). The lower the number (the more negative the number), the better the return loss. Figure 1-36 shows the formula for return loss (RL). The difference between the desired impedance and the actual impedance is the divided by the sum of their values, converted to a logarithm, and multiplied by 20.

Figure 1-37 shows a graph of return loss on a piece of RG-59 precision digital coax. This 100-ft cable is measured for return loss from 5 MHz to 3 GHz. The small spikes above the −30 dB line indicate the energy reflected back to the source at specific frequencies. In general, this cable has a typical return loss of 30 dB throughout the spectrum, which is excellent.

Figure 1-38 shows the next 100 ft of cable of the exact same reel.

There are two spikes above 2 GHz. Although both rolls of cable are typically below −30 dB, they look very different, which speaks to the nature of impedance variations and

$$RL = 20 \log \frac{\text{Difference}}{\text{Sum}}$$

Figure 1-36   Formula: return loss.

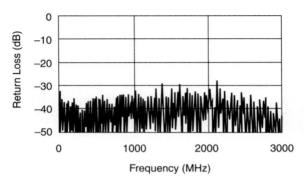

Figure 1-37 Return loss on a good cable.

return loss. Where do these spikes come from? They might be caused by a problem with a gear or wheel in the extruder, something "out of round" in the braiding process, or qual-

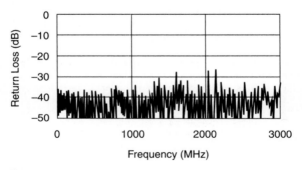

Figure 1-38 Return loss on the next 100 ft.

ity control in the extrusion of the center conductor. In other words, there's no real way to tell what the cause is.

The only solution is to make cable that is consistently good, and that consistency should include a guaranteed impedance tolerance and guaranteed return losses. All high-frequency cables have return losses. The higher the frequency, the more critical those numbers become. When installing HDTV with an uncompressed bandwidth of 750 MHz, return losses can be very important, even critical. As we analyze different systems, we will suggest acceptable return losses.

A serious percentage of return loss can be caused by the installation itself. Bending, twisting, and turning cable, especially cable with a soft foam dielectric, can dramatically affect return loss. Stepping on cables, laying equipment or heavy assemblies on cable, even how one puts on wire ties and how hard those ties are tightened affect return loss. We will consider these aspects in the installation of each type of cable.

Ultimately, return loss can affect the distance a particular signal can go on a given cable. If the cable runs only a short distance, perhaps the installer can afford to use lower quality, less-expensive cable. The problem is that this cable is a bad choice for those long runs. Do you want to carry two kinds of cable, possibly with two connectors (not to mention strip and crimp tools)? It makes more sense to standardize on one cable and make it the best performing cable you can find.

Crush Resistance

Tables 1-18 and 1-19 show how the crush resistance of a cable is directly related to foam density, impedance variations, and ultimately return loss. The ability of a cable to maintain its shape and, hence, maintain its impedance, will provide the maximum signal level for the maximum distance with the minimum return loss.

The only industry that has crush resistant specifications is the CATV/broadband industry. The Society of Cable Telecommunications Engineers (SCTE) has established parameters for crush testing of coaxes. The standard SCTE crush test uses a 4-inch flat steel plate. The plate is placed over a cable and pressure is applied at 0.2 inches per minute. The weight is increased until a 3-Ω change in impedance is noted as measured on a time-domain reflectometer (TDR).

This test is now being applied to digital and high-definition coaxes. Table 1-18 lists precision HD coax cables in the RG-59 size, and Table 1-19 lists precision HD coax cables in the RG-6 size. They show the effectiveness of high-density hard-cell foam and how it can change according to the manufacturer.

TABLE 1-18   Precision RG-59 and Crush Resistance

Precision RG-59	Average Crush Resistance
Manufacturer X	275 lbs
Manufacturer Y	164 lbs

TABLE 1-19   Precision RG-6 and Crush Resistance

Precision RG-6	Average Crush Resistance
Manufacturer X	405 lbs
Manufacturer Y	206 lbs
Manufacturer Z	213 lbs

Structural Return Loss versus Return Loss

Some manufacturers who make precision video cables for digital applications also provide measurements of structural return loss. This is a test often done with CATV/broadband coaxes. Although it is a minimal quality test and might be suitable for CATV/broadband applications, it should be considered inappropriate for broadcast installations such as SDI or HD coaxes installed in television stations.

With structural return loss, the test equipment is attached to the cable. The equipment is then adjusted to match the characteristic impedance of the cable. The impedance variations are then read and recorded. In contrast, return loss requires setting the equipment to the desired impedance of the cable. The cable is then attached and all variations from the desired impedance are read and displayed.

Note the key difference. In return loss, the desired impedance is set before the cable is attached. In structural return loss, the equipment is matched to the impedance of the cable. Because one cannot adjust the broadcast equipment to match the cable, structural return loss is not a "real-world" test and reveals only gross errors on cables. Only return loss provides a real-world test, as if the cable were attached to an installed piece of equipment.

Return Loss and VSWR

Broadcast engineers concerned with return loss will recognize the term "VSWR," commonly pronounced viz-war, which stands for *voltage standing wave ratio*. This is the same as return loss expressed as a ratio. Table 1-20 shows a range of return loss values and their VSWR equivalents.

TABLE 1-20   Return Loss and VSWR

Return Loss	VSWR
−5 dB	3.57:1
−10 dB	1.925:1
−15 dB	1.433:1
−20 dB	1.222:1
−25 dB	1.11:1
−30 dB	1.06:1
−35 dB	1.03:1
−40 dB	1.02:1

2

Constructions

Single Conductors

Single conductor wires are used to carry signals for short distances inside equipment. For those applications, these wires are called "hook-up wires." With high-frequency applications, twisted pairs or coaxes often are used inside equipment. Single conductor cable, by its very nature, has no specified impedance or capacitance. It does have inductance, and it certainly has resistance. The resistance is based on the size or gage of the wire. AWG is the standard in North America for wire size, and there are metric gage systems in other countries. Table 2-1 shows the most common gages of wire, with approximate diameters in inches and resistance in ohms per 1,000 ft. A more complete chart can be found in the Addendum.

When combining wires, you can determine the resulting gage by adding up their areas or, more precisely, their circular mil areas (CMA); as shown in Table 2-2. A simpler, if cruder, way is moving down three gage sizes when two wires of a given gage are combined. For instance, two 20 AWG wires are 17 AWG when combined, and two 30 AWG wires are 27 AWG when combined.

TABLE 2-1   Gage, Diameter, and Resistance

AWG	Approximate Diameter (inches)		Resistance (Ω)/1,000 ft at 68°F	
	Solid	Stranded	Solid	Stranded
30	0.0100	0.012	103.2	112
28	0.0126	0.015	64.9	70.7
26	0.0159	0.020	40.81	44.4
24	0.0201	0.024	25.67	27.7
22	0.0253	0.030	16.14	17.5
20	0.0320	0.037	10.15	10.9
18	0.0403	0.048	6.385	6.92
16	0.0508	0.059	4.016	4.35
14	0.0641	0.075	2.525	2.73
12	0.0808	0.095	1.588	1.71
10	0.1019	0.118	.9989	1.08

TABLE 2-2   Gage and Circular Mil Area

AWG	CMA
30	100
28	159
26	253
24	404
22	640
20	1020
18	1620
16	2580
14	4110
12	6530
10	10380

When combining different gages, or many wires of a single gage, simply add the CMAs of each wire. For instance, a 24-AWG wire and a 20-AWG wire (404 and 1,020 CMA) add up to a CMA of 1,424. That is slightly less than 18 AWG. If you had ten wires of 26 AWG (253 CMA), the total would be 2,530 CMA, slightly under 16 AWG.

Single wires can be bare or insulated. Bare wire can be solid or stranded, bare copper, copper coated with silver, or copper-clad steel. Large-gage wire, especially solid bare wire of large gage, is sometimes referred to as "buss bar." It is often used internally in equipment as a collection point for all the grounds in a design. One application that uses single wires extensively is grounding. This will be dealt with in Chapter 8.

Single wires can be insulated by many types of insulation. Table 2-3 lists the most common.

TABLE 2-3   Insulation Properties

Insulation	Symbol	Properties
Polyvinyl chloride	PVC	Low cost, flexible, colors
Polypropylene	PP	Moderate cost
Polyethylene	PE	Moderate cost, rugged
Teflon®	TFE, FEP	Very rugged, high temperature
Rubber		Rugged and flexible
Silicone		Flexible, wide temperature range
Hypalon®		Moderate cost, rugged
Neoprene®		Moderate cost, rugged

In other cable types, many of these compounds are used most often as jacket compounds because their electronic performance, such as dielectric constant, is not very good. This can be seen in Table 2-4.

TABLE 2-4   Insulation Performance

Insulation	Symbol	Electrical Performance
Polyvinyl chloride	PVC	Poor
Polypropylene	PP	Good to Very Good
Polyethylene	PE	Good to Excellent
Teflon	TFE, FEP	Very Good to Excellent
Rubber		Poor
Silicone		Poor to Good
Hypalon		Poor
Neoprene		Poor

Multiple Conductors

Multiple conductors, or multiconductor cables, most often carry control signals such as DC on and off signals. Remote controls, status lights, control relays, and similar circuits might employ multiple conductors. These signals often are described as "contact closures," indicating that there is no "signal" present, just a DC voltage that is either on or off.

Multiple conductor cables have capacitance and impedance, but these vary greatly because the positions of any two wires are not firmly established and they can move in relation to each other. Two wires in a cable do not make a twisted pair. In fact, the twisting of multiconductor cable is a random event and often used to hold the bundle of wires together. No arrangement or specific distance between conductors is intended. Of course, like single conductor cable, each wire can have resistance and inductance.

These cables are not to be confused with multiple twisted pairs or multiple coaxes, which are designed for carrying multiple signals.

Twisted Pairs

Most of this book describes two cable constructions, twisted pairs and coaxes. To understand the advantages and disadvantages of twisted pairs, we need to start with an understanding of balanced lines.

A Quick Course on Balanced Lines

Figure 2-1 shows a battery and a light bulb. A battery uses chemicals to produce electricity. A light bulb turns electricity into heat and, most important, light. But how do you get the electricity to the light bulb?

If you can conduct the electricity from the battery to the light bulb, you will have light. Because there are a positive and negative terminal on the battery and two connections on the light bulb, attaching the positive terminal to one connection and the negative terminal to the other connection, the electricity made by the battery will flow through the filament in the bulb and it will light up. If you make two thin wires out of metal, which is a conductor of electricity, and attach them to the battery and light bulb, the electricity flows as shown in Figure 2-2.

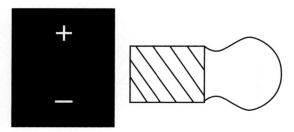

Figure 2-1 Circuit components.

The electricity flows from the negative terminal of the battery, through the light bulb, and back to the positive terminal of the battery. This is called a *circuit* from the Latin word for *circle*. There is a "circle of electricity." This "circuit" means that the electricity travels in one direction on one wire and the opposite direction on the other wire. This action defines the balanced line.

The wires could be twisted together, as shown in Figure 2-3, to make them easier to run, but the wires would have to be insulated from each other. If they were not insulated, then the two wires would touch. Electricity takes the path of least resistance, so it would flow where the wires touch instead of through the light bulb, which has much greater resistance. Therefore, the circuit would be shorter than what was intended, in other words, a "short circuit"! This short circuit prevents a light bulb from lighting up. Such a "short" in a communications circuit prevents the signal from reaching its destination.

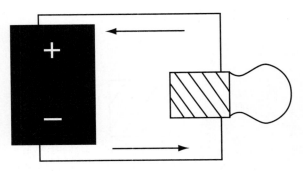

Figure 2-2 A simple circuit.

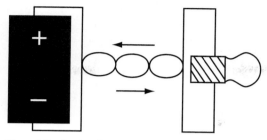

Figure 2-3   A twisted-pair circuit.

Even though the wires have been twisted together, nothing has really changed. The electrons leave the negative terminal, go through the light bulb, and return to the positive terminal. However, we can say a few additional things about the twisted pair. Pick any point along the pair and check the voltage on each wire at that point—the two voltages will add up to zero. Because one voltage is going one way and the other going the other way, they have exactly the same intensity but opposite polarities.

This fact applies to the diagram shown in to Figure 2-4, where the source of the voltage is now a microphone and the destination is a preamp (or similar device).

THE TRUTH ABOUT BALANCED LINES

Of course, balanced lines are a bit more complicated than outlined in this little book. You could write a whole book about balanced lines. A more accurate definition of balanced lines comes from Bill Whitlock of Jensen Transformers: "A balanced circuit is a two-conductor circuit in which both conductors and all

circuits connected to them have the same impedances with respect to ground and to all other conductors. The purpose of balancing is to make the
noise pickup equal in both conductors, in which case
it will be a common mode signal that can be made to
cancel out in the load. Signal symmetry, or lack
thereof, will have an effect on headroom and
crosstalk, but not on noise rejection."

A microphone converts acoustical energy into electrical
energy. The preamp/amp/speaker converts electrical energy
into acoustical energy. Therefore, what travels down the
twisted pair is a representation of the acoustical energy that
struck the diaphragm of the microphone. If that microphone
were inside a piano and someone hit middle A (440 Hz), then
the electrical flow on the wire would reverse 440 times each
second. If this were a perfect microphone, preamp/amp/speak-
er, and a perfect twisted pair, then the sound coming out of
the speaker would sound exactly like the sound that entered
the microphone.

Here is the key to the balanced line. If you chose any
place along the twisted pair, at any instant, and you could
see the signals on each wire and measure them, they would
add up to zero—identical signals, opposite polarity.

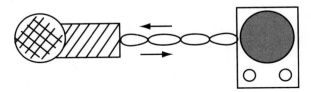

Figure 2-4   An audio twisted pair.

The transformer inside the microphone and the balanced input to the preamp makes this all work. So the circuit could be redrawn more accurately, as shown in Figure 2-5.

A transformer consists of two coils of wire surrounding a core of iron plates. The two black lines between the coils represent the plates. The coils are actually wound around the plates. The iron is much better than air at relaying the magnetic field.

Electricity traveling down a wire induces an electromagnetic field. It is so small and weak that, in the case of wire and cable, it is ignored. But coil the wire around iron, which is much better than air at passing magnetic fields, and you can induce a signal from one coil to the other. Even though the signal is passing between the coils, they do not touch. Because transformers can function without an outside power source, this is more accurately a *passive balanced line*.

There are also circuits that mimic a transformer but require power to run. These are called *active balanced lines*. There are many schools of thought regarding active versus passive balanced lines, which we will discuss.

Common Mode Noise Rejection

Balanced lines are used for analog or digital audio, premise/data computer cables, industrial control systems, and many other applications. The advantage to a balanced line is an effect called *common mode noise rejection* measured in deci-

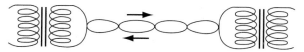

Figure 2-5   A transformer.

bels. Because it is a ratio of common signal (noise) to differential signal (the desired signal) the pair is said to have a common mode rejection ratio, or CMRR. Figure 2-5 shows how this works.

Noise is ever-present in our world. It can come from an infinite number of sources. Some of the most common include fluorescent lights and lighting ballasts, lighting dimmers (especially older SCR dimmers), power cables, motors, two-way radios, cell phones, and broadcast transmitters. Even the sun is a source of interference! Shielding the twisted pair can stop some of this interference, but some of it will get through and affect the desired signal.

When an electromagnetic noise strikes a wire, the noise induces a voltage on the two wires, as can be seen in Figure 2-6.

The difference between the signal in Figure 2-5 and the noise in Figure 2-6 is clear. The signal arrows are in different directions thus producing a "differential" signal. The noise arrows above are in the same direction, producing "common mode noise." Because the noise that strikes the two wires of the twisted pair is the same noise, it induces identical signals on both wires. These induced noise signals are the same intensity and polarity, unlike the desired signal, which has the opposite polarity.

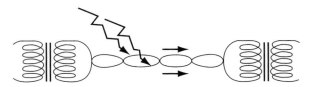

Figure 2-6   A balanced line and noise.

The noise then travels down the twisted pair until the two voltages get to the transformer, or the active balanced circuit at either end. There the two voltages meet each other and cancel each other out. Two signals traveling in the same, or common, direction are known as "common mode noise" and the process of having it cancel out is called "common-mode noise rejection."

Because the desired signal moves in opposite, or different, directions, it is called a *differential signal*. Balanced lines, or other balanced devices, often are referred to as "differential."

The better the twisted pair is constructed, the greater the common mode noise rejection will be. So what makes a perfect pair, and how can we approach such a design? Anything that affects the balance of the balanced pair will affect noise rejection. Table 2-5 lists some of those requirements and how they affect noise rejection. All of these are common problems in the design and manufacture of twisted pairs, and there are various solutions for each of them. Table 2-6 lists some of the possible solutions for each problem.

Test and measurement during the construction process can prevent many of these problems. Cable can be tested and measured to verify performance after the manufacturing is complete. Unfortunately, many of these problems have superficial remedies, but verification testing will reveal that true performance is lacking. Therefore, especially with high-frequency or high data-rate installations, performance verification after installation is strongly recommended. We'll discuss some of the ways to test and verify in Chapter 7.

TABLE 2-5A Perfect Twisted Pair

Parameter	Perfection	Imperfection	Effect on Noise Rejection
Conductor size or gage	Both wires have identical size or gage	There are variations in size or gage	Difference in size or gage means that one of the signals arriving at the source or destination device is larger or smaller in magnitude than the other and thus not identical. Cancellation will not be complete and the leftover noise signal will travel on with the desired signal.
Conductor length	Both wires have identical lengths	One wire is longer or shorter than the other due to improper twisting or cabling	The noise on the two wires will not arrive at the same time. This phase difference prevents ideal cancellation. The leftover noise signal will travel on with the desired signal.
Conductor distance	Both wires will be in exactly the same place	The wires are separated or have varied spacing	If the wires are apart, the noise signal will hit them at different times, inducing a noise signal at different times and different magnitudes. These signals will not cancel out at the source or destination. The leftover noise signal will travel on with the desired signal.

TABLE 2-6   Twisted-Pair Problems and Solutions

Problem	Solution
Variation in gage or size of conductors	Quality control when conductors are drawn Precise measurement of conductors
Variation in length of each conductor	Twist wires tightly together Be sure there is equal tension on each wire when twisting them together
Variation in distance between conductors	Twist wires tightly together Pack nonconductive materials or "fillers" around pairs to prevent them from moving Bond wires together so they cannot separate by themselves

Multipair Snake Cable

Twisted pairs often are assembled in groups. These cables are known as *multipair* cables. When they carry audio signal, they often are called *snake* cable. In Europe, they sometimes are called *multicore* cables. Data versions sometimes employ unshielded twisted pairs. Audio cables more commonly use shielded twisted pairs. Shielded and unshielded cables will be discussed in a later section.

One advantage to using multipair snake cable is being able to install one large cable instead of many smaller ones. This can be especially cost effective for very long runs, where labor is a significant cost factor. Snake cable tangles less than individual cables and has a neater appearance. We'll discuss just how to install snake in Chapter 7.

Coaxial Cable

Figure 2-7 shows a cross-section of a coax. If you draw a line through a cable of this construction, you will see that it dissects all the elements of the cable. In other words, all elements are on the same axis, or coaxial. This construction has one immediate advantage. The center conductor, dielectric, shield, and jacket are "locked" together and cannot move, unlike a twisted pair, where the pair can move and separate. Therefore, the variation of impedance, or "impedance tolerance," of a coax is very small. Even when the cable is bent, flexed, or stepped on, a good quality coax is quite resilient and will maintain its designed impedance much better than a twisted pair.

This is not to say that a coax cannot be damaged or change its impedance. Rather, a coax maintains its performance better than many other types of cable. For this reason,

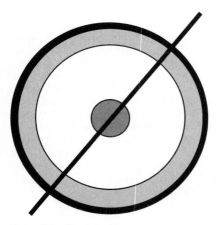

Figure 2-7   Coaxial cable.

coax has been the cable of choice for many high-frequency applications. It is also easier to place very large conductors in coaxes. This allows it to be used for high-current applications such as sending power to antennas for transmission. There are other construction requirements, such as designing the cable for the correct impedance.

Coax and Impedance

Figure 2-8 shows the performance of coaxes of different impedances at different frequencies. Each parameter has a different characteristic performance curve. For instance, if you are sending signals down a coax and want the least attenuation (signal loss), you use a 75 Ω coax. If you are sending high power (high voltage and high amperage), you would use 50 Ω coax, which is a compromise between 30 Ω (power) and 60 Ω (voltage). Many other coaxes have been designed with other impedances, such as the Arcnet® cable at 93 Ω. This slow-speed computer cable trades attenuation for low capacitance.

THE LEAST LOSS

Actually, the least loss in a coax falls closer to 77 Ω. Because the impedance calls out the ratio of dimensions between the cable elements, 77 Ω works out to a very odd ratio. Moving it to 75 Ω, which only slightly affects attenuation, allows the use of standard size components. This also applies to 51.5 Ω and 52 Ω coaxes, which still exist. Most of these high-power transmission cables have been standardized to 50 Ω.

In most coaxes, the capacitance is preordained by the choice of dielectric and the spacing necessary for a specific

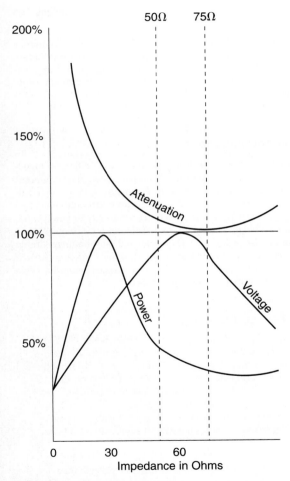

Figure 2-8 Coax impedance and loss.

84

impedance. Therefore, a coax designed for a very low capacitance would have a very high impedance, and one designed for a very low impedance would have high capacitance. In actuality, it is hard to manufacture a coax with very low impedance. The ratio of sizes also becomes increasingly low and hard to construct.

If a coax is being used as an analog audio interconnect, the frequencies are so low that it wouldn't make any difference what the impedance is (see Table 1-13). Therefore, an analog audio cable could be designed for the minimum capacitance. The limitations would be the smallest center conductor one could work with and the largest diameter one could accept.

A very large diameter, say 0.400 inches, and a very small center conductor, say 30 AWG, would produce a cable with a capacitance of just under 7 pF/ft and, if you were wondering, an impedance of 173 Ω. The small center conductor with such a huge core would be extremely difficult to manufacture and very hard to install connectors.

Table 2-7 lists a number of coaxes commonly made, in order of capacitance. It should be noted that many of these are not the most common constructions and may be hard to obtain.

There are variations on each of the types listed in Table 2-7. This information shown applies to the most generic version of each type. You also can find plenum-rated versions, which contain Teflon cores and sometimes a Teflon® jacket, or direct-burial or outside versions that are more rugged. Plenum versions may have slightly less capacitance than the non-plenum versions. These different versions may have different diameters than the generic versions listed in Table 2-7.

Multicoax Cables

Multicoax cables have, until recently, been used only in a few specialized applications such as RGB cable. RGB con-

TABLE 2-7 Low Capacitance Coax for Analog Audio

RG Type	Capacitance pF/ft	Impedance Ω	Diameter (in)	Notes
RG-63	9.7	125	.405	
RG-62	13.5	93	.242	Arcnet data cable
RG-71	13.5	93	.245	
RG-180	15.4	95	.102	
RG-6	16.2	75	.275	Nitrogen gas-injected foam
RG-59	16.5	75	.235	Nitrogen gas-injected foam

tains three or more identical coaxes most often color coded with red, green, and blue jackets. They carry the red, green, and blue components of a color video picture. Although common for many years in broadcasting, RGB is appearing on some high-end consumer television equipment, and suitable RGB cable must be used. Synchronization, or "sync," signals, which often are required to get the highest level of performance, require up to four or five coaxes, sometimes even more than that. There is no standard for the color code of coaxes aside from the basic red–green–blue, so other colors can be used, although yellow and white are often used for those other signals.

The second application for multiple coaxes is for computer monitors, computer projectors, and similar systems. These are driven by a signal called VGA, for *video graphic*. There are newer systems that employ S-VGA, for *super video graphic*, or X-VGA, for *extended video graphic*. Some of these employ coaxes and twisted pairs in one cable. We will look at these in more detail later on.

The last application for multiple coaxes is ENG/EFP, which stands for *electronic news gathering* and *electronic field production*. You know—those trucks, both small and large that go around shooting video for your local station, the networks, or even higher-quality television shows and commercials. The snake cables, or umbilicals, from the van often have multiple video coaxes, sometimes with multiple audio twisted pairs. We'll be looking at them in detail later on.

Unbalanced Lines

Many consumer and nonprofessional devices use cables that are unbalanced. This translates to a significant saving because active or passive balancing devices at the source or destination are not required. However, the noise rejection offered by a balanced line is also compromised.

Unbalanced cables have two variations. The first is a twisted pair (Figure 2-9) and the second is a coax-like design (Figure 2-10).

BALANCED SPEAKERS?

Couldn't you build an amplifier with a balanced output? Of course you could, but it would probably be an exercise in frustration. The output of most amplifiers is very large, often many amps of current. To be able to interfere with such a huge signal, noise would have to be a significant percentage of that huge signal. Luckily, most noise is in the millivolt (thousandth of a volt) or microvolt (millionth of a volt) region. Noise that small will have an imperceptible effect on speaker cables. You also would have to build a balanced input speaker. Most crossover inputs on speakers are not balanced.

How can a twisted pair not be a balanced line? If the signals on the two wires are not equal and identical, then, by defini-

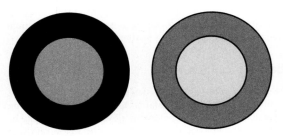

Figure 2-9   Twisted-pair unbalanced line.

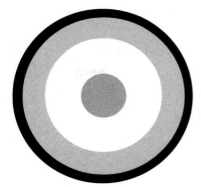

Figure 2-10 Coax unbalanced line.

tion, it is not a balanced line. The most common type of paired cable run in unbalanced mode is speaker cable. One wire is attached to the positive output terminal of the amplifier and the other is attached to the ground. *Ground* is defined as a place of zero potential. If this were a balanced line it would have an identical value to the positive terminal but the opposite polarity. This is not the case with speaker cable.

Therefore, speaker cables are not balanced lines. And no matter how beautifully the two conductors might be twisted together, there is no common-mode noise rejection accomplished. There are other signal types, such as RS-232 control signals, which run unbalanced on twisted pairs. The data rate for these systems is low, and the distances they go are not very far, so they generally work well.

The other unbalanced cable is a single conductor with a shield, shown in Figure 2-7. Although the design looks like coax and you can use coax for this application, most of these cables lack the performance parameters of quality coax.

For one thing, most of these cables are made entirely with PVC. This gives them very high capacitance. However, they rarely run more than 10 ft (3 m), so this is not a problem. They often are made with aluminum conductors in both the center and the surrounding shield. This reveals their true "low-cost" construction because the resistance of aluminum is considerably worse than that of copper. The short distances these cables run also make this a minor consideration. The spiral shield these cables often use was discussed on pages 19–20.

Like the unbalanced twisted pair, the center conductor is the "positive" voltage of the signal, and the shield is the ground. The two signals are not equal and opposite, so this is not a balanced line.

If you tried to run such a cable as a balanced line, it would be a very poor choice because it already breaks some of the requirement necessary for a quality balanced line. First, the two conductors are different sizes, with the AWG equivalent of the braid often being many times larger and with much lower resistance than the center conductor. Second, the two conductors are kept apart by the dielectric.

NEC Fire Ratings

The National Fire Protection Agency (NFPA) publishes the National Electrical Code (NEC) or NFPA-70. It outlines how to install systems, such as audio and video systems, as safely as possible. You can get a copy of the NEC at any good technical bookstore or call the NFPA at 1-800-344-3555 or 1-617-770-3000.

The NFPA is not a governmental organization but many communities use it as their basis for fire safety. It is therefore incumbent upon you, the installer, to determine whether your fire marshall, building inspector, electrical inspector, planning commission, or board of permit appeals subscribes to the NEC. If they do, then so must you!

Keep up to date with the NEC. There are changes, additions, and deletions in every new issue. One of the key things the NEC requires is the testing of cables to determine their fire safety. The details for this can be found in the NEC section listed in Table 2-8. More detail on the NEC code can be found in the Addendum, including a substitution chart.

TABLE 2-8   NEC Articles

NEC Article	Subject	Details
725	CL2	Class 2 cables, power limited but no voltage rating
	CL3	Class 3 cables, power limited with 300-volt rating
	PLTC	Power-Limited Tray Cable, a CL3-type UV and moisture-resistant cable for use outdoors; must pass the vertical tray flame test
760	FPL	Power-limited fire-protective signaling circuit cable
770	OFC	Fiber optic cable with conductive metallic elements
	OFN	Fiber optic cable that is entirely non-conductive
800	CM	Communications cable
820	CATV	Community antenna television and radio distribution

The most common NEC ratings for audio and video installers are CM, CMR (riser), and CMP (plenum). You also might use CL2, CL2R, and CL2P, as well as CL3, CL3R, and CL3P.

Two Stories About Fire

There were a number of incidents that motivated the NFPA to write the NEC. Here are two of these stories.

The first was a fire on February 27, 1975, in the AT&T 11-story Central Office Switching Terminal at 13th Street and 2nd Avenue in New York City. The fire started at 12:25 AM in the cable vault under the building and burned for 16 hours. Although there were no fatalities, the toxic fumes produced by the burning wire sent 175 firefighters to the hospital.

The fire put 170,000 phones out of order in a 300-block area of the city. It required 1.2 billion feet of new service wire and 8.6 million feet of cross-connect wire. It took a 4,000-man Bell System task force 22 straight days, working in 12-hour shifts, to restore service, and a total of 562 person-years to rebuild the facility.

It is interesting to note that 51 percent of the 700 firefighters who fought the blaze never made it to the 20-year mandatory retirement and that 57 of them (8 percent) got colon cancer.

The second incident was the MGM Grand Hotel fire in Las Vegas at 7:10 AM on November 21, 1980. There was a small fire in the kitchen of a closed deli restaurant caused by the overheating of a compressor that cooled the pie storage rack. The fire quickly spread to the wooden ceiling tile and the vinyl and polyurethane-cushioned booths.

Soon the entire restaurant reached flashover, at which time balls of flame were spreading to the casino area. Gamblers refused to leave the tables or were prevented from picking up their bets by the pit bosses until they were overcome by toxic fumes.

Soon the fire spread to the highly flammable, clear-plastic suspended ceiling in the casino. The eventual stampede to the exits prevented many from getting outside. Seven people died in the casino.

The toxic smoke was drawn up the elevator shafts, up special earthquake ducting, and into the plenum drop-ceiling, where it was pulled out by the air conditioning system and introduced to the rest of the hotel.

Twenty-two people died in the hallways and stairways. Twenty-five more people died in their rooms. The air vents had successfully scrubbed the soot from the smoke, leaving only the invisible toxic fumes. People only needed to block their doors or air conditioning vents with wet towels or work their way to the roof.

Sixty-five bodies were found above the first floor, the actual location of the fire. A total of 85 people died, and 600 more were injured. The hotel had passed a fire inspection 6 months previously.

You can understand the need for fire ratings and plenum ratings in particular. Although all cable will burn given enough heat, plenum cable has been manufactured and tested to be a minimal fuel source for and to resist spreading the fire. The plenum test is called the Steiner Tunnel Test. Cable manufacturers send their cables to third-party test facilities, such as the Underwriter's Laboratory, or ETL, to verify the fire safety performance of each cable design.

3

Analog Audio

The word analog means *copy* in Greek. Ana means "according to" and logos means "relationship." We often speak of something being "like" or "analogous" to something else. Analog audio is a signal created "according to the relationship" of something in the real world, that is, a sound and a copy of that sound.

Let's look at a piano for an example. It contains strings that are drawn tight and tuned to various frequencies. Let's say we hit the key commonly known as middle A. By pressing the key, we make a felt-tipped hammer hit a certain string that will vibrate 440 times per second. We say that the string is vibrating at a frequency of 440 Hz. (Yes, I know, some piano purists will say "middle A" is 436 Hz.)

A graph of the movement of the string (Figure 3-1) shows two of the 440 cycles the string would move each second.

If the piano were floating in outer space, we would see the string move, but we would hear nothing because there is no air. Here on earth, the vibrating string causes the air around it to compress and expand at the same rate as its motion. If we graphed the air pressure anywhere around the string, it would look like Figure 3-1.

But now our 440-Hz tone reaches our ears, where our eardrums are moved back and forth, allowing us to hear it.

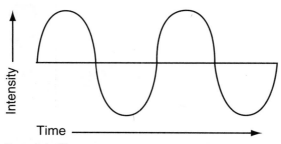

Figure 3-1   Sine wave.

If we graphed the movement of our eardrums, or any other moving part of our ears, it would look like...Figure 3-1.

If we substituted a microphone in place of our ears, the diaphragm (moving part) of the microphone also would respond to the air pressure caused by the vibration of the string. If we graphed that movement, it would look like...Figure 3-1. In fact, a microphone is really a converter. It converts acoustical energy (air pressure) into electrical energy (audio signal). The signal leaving the microphone, going down the twisted pair, looks exactly like...Figure 3-1.

TELEPHONE TWISTED PAIRS

Until the dawn of the digital era, all telephones ran on analog twisted pairs. The system impedance, sources, and destinations, for all telephone equipment is 600 Ω. In truth, their voice pairs are 900 Ω. But their twisted pair cable isn't 600 Ω, 900 Ω, or anything even close. In fact, it is much lower, probably 30 to 70 Ω, and the actual value changes with frequency. The phone company is one of the few entities whose twisted pairs do indeed run a mile or more. In fact, the maximum distance from any customer to the central office is 13,000 ft, more than 2 miles. So how do they make it work? By "loading" the line and adding extra inductance to balance out the capacitance. For their top-of-the line customers (broadcasters) they use 111C coils that reduce 600 Ω to 150 Ω to extend the distance and simplify equalization.

And every following part, if intended for analog audio, will have an identical waveform going in and going out. It might be bigger, it might be smaller, but it will be the same. Any variation from the original would be undesirable distortion or noise. Eventually, the electrical signal travels through an amplifier to a speaker. The motion of the cone could be graphed like...Figure 3-1. If this chain were perfect, that reproduced sound would be absolutely identical to the original and could be graphed like...Figure 3-1.

Of course, there is no perfect microphone, cable, preamp, console, mixer, amplifier, speaker, or anything else. All of them add noise and distortion. Some devices, such as equalizers, compressors, limiters, or expanders, even manipulate the sound in ways different from the original. If compared to the original, these sounds would be considered noise or distortion, but they sound just fine to our ears. This shows how imperfect our ears really are. A distorted sound often can be more pleasing than the original.

The whole point is that each step is a copy, an analog, of the previous step, so such a system is called *analog audio*. Cables made to run analog audio have a number of specific features. As previously mentioned, the impedance of the cable at audio frequencies usually is not an issue. Unless your cable is a mile long, it doesn't matter what its impedance is.

But impedance is important. Not the impedance of the cable, but the impedance of the system. We're really talking about two impedances, the impedance of the source and the impedance of the destination. In the early days of the telephone, both were 600 Ω, and all devices that went into or out of the telephone company had a system impedance of 600 Ω. Professional audio gear followed suit, because much of it was made by Western Electric, the supplier to the Bell System.

It's a different story now. The distance a signal can go is based on the relation between the source impedance (the lower the better), the destination impedance (the higher the better), and the capacitance of the cable in between. To determine just how far you can go, you must compare impedance to capacitance or, more accurately, impedance to capacitive reactance.

Capacitance versus Impedance

When cables approach a quarter of a wavelength, it is important to choose an impedance and match the cable to that impedance. Some use an eighth of a wavelength, or even a tenth of a wavelength. This would make the quarter-wave distances shown in this book even shorter and the arguments about cable length even more critical.

For frequencies below the megahertz range, such as analog audio, the impedance of the cable doesn't make a difference. But the impedances of the source and destination devices are important. Although the cable impedance doesn't matter, the capacitance does. The interaction between the system impedance, specifically the source impedance, and the capacitance of the cable, will determine just how far you can go.

Table 3-1 shows the distance one can go based on the source impedance and the capacitance of the cable. This applies to balanced or unbalanced systems.

TABLE 3-1 Capacitance versus Impedance

Source Impedance	−1 dBm at 20 kHz			
	15 pF/ft	20 pF/ft	30 pF/ft	50 pF/ft
50 Ω	5,406 ft	4,055 ft	2,703 ft	1,622 ft
150 Ω	1,873 ft	1,352 ft	901 ft	541 ft
600 Ω	451 ft	338 ft	225 ft	135 ft
1,000 Ω	271 ft	203	135 ft	81 ft
10 kΩ	27 ft	20 ft	14 ft	8 ft
50 kΩ	5.4 ft	4 ft	2.7 ft	1.6 ft

Table 3-1 shows examples of capacitance and capacitive reactance. The frequency chosen is 20 kHz, generally regarded as the highest audible frequency. If you disagree with the choice of 20 kHz, you can pick any other frequency and do the calculations yourself. If you wish to compare other source impedance values or other capacitance values, insert them and calculate the results.

To simplify this analysis, the calculations ignore resistance. When the capacitive reactance equals the source impedance, (left column), then the signal will decrease by half (−3 dBm). The distances shown in Table 3-1, are equivalent to a 1 dB loss, arrived at by multiplying the −3 dB distance by a factor of 0.51.

Systems with high source impedance, such as analog consumer audio equipment, generally around 10 kΩ, cannot send these signal very far down cables. Even with ultra-low capacitance cable, distances beyond 28 ft can produce measurable, and audible, high-frequency loss.

High-end consumer manufacturers are well aware of this and much of the equipment has output impedances below 10 kΩ, some as low as 1 kΩ, or even lower. But how do you know what the output impedance of your device really is? It is not written on the unit. It may not even be in the owner's manual. A manufacturer might say "10 kΩ" because it has determined that the load impedance of the next device should be no lower than this value. The manufacturer's Web site might offer some information or a phone call might be in order.

However, if these systems are designed or converted to a lower impedance, the distances these signals can go without significant loss can be greatly extended. For equipment with high-output impedance, we'll talk about changing and matching impedances on page 122.

Very high-impedance systems (50 kΩ) such as electric guitars are very limited by distance. The highest note on a guitar is generally around 2 kHz, which allows such a high impedance to be used. Some of the distinctive electric guitar sound might be due to this filtering effect. Certainly much of the harmonics of the music played is filtered out.

What's a Decibel?

A decibel is a tenth of a bel. OK, so what's a bel? It's a unit of relative level. There are many different decibels: dBv is a voltage ratio compared to 1 volt, dBm refers to a standard level of 1 milliwatt, dBu is referenced to 0.775 volts, dBk refers to 1,000 watts, and a dBf refers to 1 femtowatt (10^{-15} watts) used in RF measurements. A decibel without description is a relative term. So, the 100-watt amplifier in Table 3-2 is equivalent to +50 dBm (reference 1 mW), −10 dBk (reference 1 kW), and +20 dBw (reference 1Ω). My friend, Bill Ruck, uses dBhp (dB above 1 horse power) to point out that there are lots of "dBs."

On the other end, microphones generally have output impedances near 150 Ω. This can be a dramatic difference in

terms of distance and the reason why full-range microphones can easily be installed with cables a hundred, even a
thousand, feet long and still provide excellent performance.
There are even some microphones that allow the user to
select the output impedance, some as low as 50 Ω. This dramatically affects distance. Because the capacitance has so
little effect at these low impedances, microphone response is
extremely flat even over long distances.

These effects apply to balanced and unbalanced systems.
Of course, professional balanced systems are almost automatically low impedance because of the distance advantage.
Home theater and consumer audio use an altogether different system.

Signal Level

There's also the question of level, which is the actual amount
of signal on a cable. Although there are two key levels for
audio, there are a number of signal levels, or intensities,
that might be applicable. Table 3-2 shows a number of common signal levels and where you might encounter them.

TABLE 3-2   Signal Level

Signal Level	Common Device
−70 dBu	Old condenser, old ribbon microphones
−60 dBu	Common "lowest" mic output
−50 dBu	Standard microphone output level
−30 dBu	Electric guitar output
−10 dBv	Standard level in home hi-fi/consumer gear
0 dBu	Standard professional reference
+4 dBu	Output of most pro analog audio equipment
+8 dBu	Some hi-end professionals prefer this level
+50 dBm	Output of a 100-watt power amplifier

From the lowest level to the highest level, there is a range greater than 100 dB, a ratio of ten billion to one. It is quite amazing that signals so small and large can be easily generated and preserved relatively free from noise and distortion.

Unbalanced Analog Audio Interconnection

Most home theater, high-end, and consumer audio devices are unbalanced. They are based on the RCA, or phono, connector. What is the impedance of these connectors? Because it doesn't matter what the impedance of the cable is, it definitely doesn't matter what the impedance of a connector that is, at most, a couple of inches long.

So what constitutes a good home audio cable? As can be seen by Table 3-2, even with the lowest capacitance, you are limited to a few feet. It is no wonder that trying to run such a cable even 50 ft can exhibit significant high-frequency loss. Devices with lower output impedance can be a real help in this situation.

So what can a home/consumer installer do? There are two answers. The first is to use cables with a capacitance as low as possible. One excellent choice would be one of the newer precision video cables intended for digital video, and the reason is instantly apparent. These cables have very low capacitance. Table 3-3 lists a number of popular-size digital video cables; their capacitances are almost identical. The capacitance is locked in by the choice of dielectric and the dimensions necessary to arrive at the desired impedance.

Of course, the impedance is not a factor for analog audio, but the resulting low capacitance is important. Because the dielectric in these cables is cutting-edge nitrogen-gas–injected foam, they have capacitances 20 percent lower than similar-sized solid dielectric coaxes.

What if you wish to go more than 27 ft between devices? Then you will need to convert to a lower impedance. While you're at it, you might as well convert to a balanced line and get good noise rejection. We'll discuss both of these options in upcoming sections.

TABLE 3-3   Video Cables for Audio

Description	Diameter (in)	Capacitance
Miniature precision	0.159	16.5 (pF/ft)
RG-59 precision	0.235	16.3 (pF/ft)
RG-6 precision	0.275	16.2 (pF/ft)

Balanced Analog Audio Interconnection

If you read the section on balanced lines (pages 73–81), you know that balanced lines naturally reject noise because the pairs are differential and any noise is common mode. Therefore, it is just a question of what style of balanced line cable to use. Table 3-4 lists a number of different single-pair balanced line cables, with application suggestions. We will discuss each type in more detail.

Microphone Cable

Microphone cable, also called mic cable, is used between microphones and associated equipment. The signals running along microphone cables are among the lowest levels in the audio and video world. Mic signals can be as low as –60 dB or even –70 dB for older condenser mics or ribbon-type mics. This means that the signal traveling down a mic cable can be very weak. This also makes those signals susceptible to interference. Because electronic performance of the cable is low on the list of key parameters (Table 3-5), one might

TABLE 3-4 Balanced Line Cables

Construction	Insulation	Shielding	Jacket	pF/ft	Application
Twisted pair	Foam PE	French Braid + drain	Matte PVC	13	Microphone
Twisted pair	Rubber	Braid	EPDM	34	Microphone
Twisted pair	PVC	Braid	PVC	50	Microphone
Starquad	Polyethylene	French Braid + drain	Matte PVC	50	Microphone
Starquad	Polyethylene	Braid	Matte PVC	50	Microphone
Twisted pair	Polyethylene	Foil	PVC	25	Mic/Line level installation
Twisted pair	Polypropylene	Foil	PVC	30	Mic/Line level installation
Twisted pair	PVC	Foil	PVC	50	Mic/Line level installation

wonder why performance is not first on the list. The answer is in the length of the average microphone cable.

The key to good microphone performance is not to leave the signal at microphone level for long distances. The sooner those signals are amplified to line level (+4 dBu or more), the better they will resist external noise. In most studio installations, microphone cables rarely run more than 20 to 30 ft. In some high-end studios, if microphones run farther, a short cable is attached to a microphone preamplifier, where it is boosted to line level. At line level, those signals are much more resistant to noise.

This is not to say that you cannot run microphone cables a long distance. Table 3-1 clearly shows that with a low impedance and average capacitance in a cable, signals run 1,000 ft or more. Of course, even though the frequency response might be acceptable, you are asking for noise and interference to show up. You have just provided these elements with a 1,000-ft antenna to work on!

So you put that microphone cable in conduit, where it is well protected from noise. Why did you pay for ruggedness and flexibility and then stick the cable where those expensive features are not even used? Simply convert to a high-quality install cable, as shown at the bottom of Table 3-4.

Microphone cable has a number of properties. They are listed, by order of importance, in Table 3-5.

TABLE 3-5   Microphone Cable Properties

Order	Property	Reason
1	Ruggedness	Handled constantly
2	Flexibility	Ease of use
3	Self-noise	Quiet while in operation
4	External noise rejection	Resists external noise
5	Electronic performance	Short runs, not critical

Installed Cable

You might think that installation is an easy thing to define: Unfortunately, it is at the heart of some controversy because, if a cable is installed, it most often comes under the NEC.

HORROR STORY #1,000,000

I was working with a director of engineering at a TV facility in a state college. By coincidence, the campus straddled two different counties. So we had the fire marshal from each county and the state fire marshal telling this guy what to put in. One said, "We go by the NEC. Follow that." The next one said, "We don't follow anything. Do what you want." And the third guy, who was stuck in the dark ages, said, "Everything you put in must be 100 percent Teflon®." This argument effectively shut down the project for 6 months. They finally agreed to go by the NEC.

Most professionals would concede that a cable is installed whether it is hidden behind a wall, running in a conduit, on a permanent tray, between floors, or above a drop ceiling. Cables in these locations are generally difficult or impossible to get to (tray cable might be the exception). All are intended to be placed and then left. It is unlikely they will be touched again until they are being removed permanently.

It would also follow that anything outside, where the NEC fire ratings do not apply, is not installed. Of course, those cables may be just as permanent as those inside buildings but the fire danger they represent outside is not a consideration.

What about a snake cable in a theater? It's rolled out on the floor in full view. It may be there for days or weeks while the performance runs. Is that installed? What about the tri-axial cable in a TV station? Cameras are connected with triax to pedestals or plates on a wall, often permanently. The stuff in the wall is installed. Is the stuff attached to the wall installed?

I could give you my opinion on what is and isn't installed, but that wouldn't help you if the fire marshall, board of permit appeals, or building inspector (or others who carry the Big Hammer) think otherwise. Therefore, if you are planning a design or installation, I strongly suggest that you seek the opinion of the person who will approve the plans and/or installation.

Ruggedness

To be sure, it is hard to find any microphone cable that is fire rated in any way, indicating that is not intended for installation. Many of the properties of microphone cable are designed for ease of handling. If it is installed, it is by definition not handled. There are install versions of each of these cables, at a fraction of the cost, and often with better performance.

Microphone cable is out where you can see it, for example, on a stage, on a podium, or run though a mic boom or other support. It is constantly being handled and flexed, not to mention stretched, thrown, walked on, bundled, unbundled, and otherwise manhandled. This is the reason ruggedness is first on the list. You might argue that some cables are in a protected environment, such as a recording studio, but you still want them to last as long as possible. Although the abuse may be less, it can happen.

The key to ruggedness is the choice of jacket. Table 3-4 lists a number of jacket formulations. The most rugged is EPDM or artificial rubber. It is extremely tough yet surprisingly flexible.

Many jackets are made of PVC, but not all PVC is the same. There are many ruggedized versions of PVC. Some are more than 40 times more cut resistant than other softer versions. Although not as rugged as EPDM, PVC is acceptable for many standard uses.

Flexibility

Because flexibility cannot be measured (see below), it is hard to judge what is acceptable flexibility. Some cables are made to be as limp as a wet noodle. Such flexibility comes at a cost because ultra-flexible cables may be significantly less rugged than other cables, prone to fail, and replaced more often. Also, the compounds required for ultra-flexible cables are considerably more expensive than "standard" PVC.

How Flexible Is Flexible?

Want to make a name for yourself? Just invent a way to measure flexibility on a cable. This is not true of flex-life, which is the number of flexes a cable can take until it fails. There are standard tests and parameters for testing flex-life. However, there is no standard test to measure flexibility. The only way to compare cables… is to compare cables. Just get a piece of each and flex it. And if you come up with a way to test and measure flexibility, people might name the unit of flexibility after you! Good luck!

Self-noise

When a cable is moved, bent, or flexed, the various components inside the cable tend to move. Just like walking on a thick carpet on a dry day, this movement can create static charges inside the cable. These are heard as "crackling" noises when the microphone is plugged in and working and the cable is moved, flexed, or stepped on. Since these noises are generated by the cable itself, so they are called *self-noise* or *triboelectric noise*; tribo is Greek for "to rub."

Movement of the inside parts also changes the capacitance very slightly. This can be heard when the microphone signal is amplified and is one reason foil shields are not recommended for microphone cables. When flexed or stepped on, you can clearly hear the change in capacitance as pops or crackles. If the cable is permanently installed and will not be moved while in operation, foil shields are fine. If the cable will be moved while in operation, stick with serve/spiral shields or, better yet, braid or French Braid shields. You might want to review shields on page 18.

External Noise Rejection

There is also natural rejection of a balanced line, because virtually all professional microphones are twisted-pair balanced line devices. The requirements of a balanced line (page 73) apply here. Because Microphone cables are intended to be flexed, so it is especially important that the two conductors of a balanced line stay close to each other. Any variation in spacing will increase the ability of the pair to pick up noise. It is unlikely that the signal on microphone cable will radiate to other cables because the signal is so small.

Performance

Even though performance is at the bottom of the list, it cannot be ignored. Once those other requirements are taken care of, performance becomes a serious consideration, especially in critical applications, such as recording studios, where the quality of the signal is paramount. And, like all other cables, cables with the lowest capacitance will have the highest performance. One cable in Table 3-4 with a capacitance of 13 pF/ft, two to four times lower than other microphone cables. This might not be the ruggedness champ or the low-noise pickup champ, but it is by far the raw-performance champ.

There is one construction of microphone cable that deserves special attention, the starquad cable. This cable has four conductors rather than the usual two conductors for a balanced line. Figure 3-2 shows a cross-section of a starquad cable.

Wiring a starquad cable

Because you need only two wires for a balanced line, you combine conductors to make such a pair. Most starquad

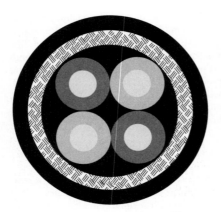

Figure 3-2   Starquad microphone cable.

cables are color coded, so you can simply combine the conductors of the same color insulation. For instance, if you have two blue wires and two white wires, you combine the blues into one wire and the two whites into one wire.

You might recall from Chapter 2 that a perfect balanced line would have both wires in exactly the same place. That way, any noise hitting the two wires will be absolutely identical on both wires and completely cancel out at the source or destination. Of course, it's impossible to put two wires in the same place...or is it? That is exactly what starquad does. By combining the wires, you are essentially "spreading out" each conductor. Because all four conductors are intertwined in a spiral, conductors are inside each other. Further, when you combine them, as shown in Figure 3-3, you can see that the resultant conductors are in the same place.

Figure 3-3 How starquad works.

However, there is no free lunch, and the payback here is in capacitance. Because you are combining conductors, you are also combining the capacitance. Therefore, capacitance on starquad cable is high, sometimes more than 50 pF/ft. High capacitance only means you can't go very far. The 50-pF distance shown in on Table 3-1 is 540 ft at 150 Ω, a common microphone impedance. Because microphone cables are rarely that long, capacitance is not a problem. If you can get microphones with even lower impedance or select the impedance internally, then you can go even farther.

Starquad cables come in a variety of sizes and constructions. Smaller cables generally indicate smaller conductors and less ruggedness. On the other hand, if you are wiring up a microphone boom in a broadcast installation, a smaller cable can be a godsend.

There is even miniature starquad cable. Because of its small size, these conductors often are not made of copper, which could easily break, but of a special bronze alloy called CT137. Bronze has higher resistance than copper, so each strand in this bronze conductor is silver plated to bring the total resistance up to the same value it would have if it were all copper. This makes such a miniature cable as strong as many full-sized cables.

When to Use Starquad

Starquad is ideal when you know there is serious interference in the area of your installation. Starquad is effective against very low-frequency noise, such as 60 Hz power and lighting. High-coverage braid shielding now carries the noise rejection well into RF frequencies.

There are four kinds of sound engineers who should definitely consider starquad cables. The one thing they have in common is that they are on the road. Often their needs in cable are identical, but it is rare that they meet, and if they did, it's doubtful they discuss cable.

60 Hz AND STARQUAD

I once went to a talk about shielding and grounding given by Neil Muncy, of Toronto, Canada. He wired up an amp with a balanced line input and then used a hand-held tape demagnetizer, wired so it was always on, as a 60 Hz generator. Neil then took 20-ft samples of different kinds of wire and attached an XLR to one end of each sample. As he wrapped each wire, one at a time, around his 60-Hz demagnetizer, after a turn or two, you could clearly hear the 60-Hz hum from the speaker. Some cable got up to five or six turns, which was even better. Until he got to starquad cable. He wrapped the starquad around and around, until the entire cable was wrapped around the demagnetizer...and not a hint of 60 Hz. I remember turning to my friend, Lowell Moulton of Sony SIC and saying, "This starquad stuff actually works!"

The first are those hearty souls who record interviews, sound effects, or bird songs. They can find themselves in the strangest places, trying vainly to pick up a pristine track. Wouldn't it be nice if they didn't think about 60 Hz and other interference? Starquad would give them one less thing to worry about.

Next are those unsung sound engineers on the road with traveling shows, musicals, and stage plays. There are lighting cables all over the place, with 60-Hz noise everywhere. Starquad would be a good choice because you don't know what you will encounter next. Each venue is different, with a different set of problems.

The next group consists of rock band sound engineers. With bands, you can be in a different club each night. These clubs often are underground, in the basement of some old warehouse, in a factory neighborhood. You'll have to put up with antiquated building wiring and lots of uncontrollable electrical noise. Starquad might make the difference between usable and unusable sound.

The fourth group consists of ENG and EFP engineers. They should definitely consider starquad because they never know where they'll be. They don't know what kind of electrical environment they'll be in. A pristine track is essential, no matter the surroundings. Starquad is the ideal choice for these applications.

Starquad Microphone Snake Cable

If you are installing a system where noise is a constant problem, you might want to consider starquad snake cables. These are multipair, or should I say multiquad, cables from 2 to 32 quads. They will maintain the low-noise characteristic of your individual starquad cables.

Starquad snake can be permanently installed, if it has the right NEC rating (see pages 90–91). The only concern is distance. If you are running multiple mic lines, use less than 540 ft. If you are running line-level signals, You can use 135 ft for 600 Ω systems (Table 3-1). In truth, most modern mixers or consoles have a very low output impedance, often 100 Ω or less. You can check the manual on that piece of equipment or call the manufacturer and find out. If you don't know, you might consider 135. If the impedance is 150 Ω or lower, then 540 ft would be a good maximum distance.

Remember, the NEC requires any installed cable to have a rating. It doesn't say what the rating must be, just that it has a fire rating. Most starquad microphone cables are intended for connection from the microphone to the floor, wall panel, or snake box, and not for installation. Starquad microphone cables therefore are unrated and cannot be legally installed if your local area uses the NEC as its guideline.

Install Cable

Standard twisted-pair analog audio install cables are constructed as shown in Table 3-6 below. These constructions are very simple, and with good reason. They will be installed and left in place. They do not have to be rugged beyond the installation itself. They do not have to be flexible. In fact, flexibility can be counterproductive in an installed cable.

TABLE 3-6   Standard Installation Cables

Gage Size	Insulation	Capacitance	Shield	Jacket
26–20 AWG 22 AWG was the standard; 24 AWG is becoming more popular	PVC Polypropylene Polyethylene	50 pF/ft 30 pF/ft 25 pF/ft	Foil Braid Most install cables are foil shielded	PVC Rarely is any other plastic used

Flexible or Stiff?

Flexible cables can be nice and limp, which makes them easier to coil and uncoil. This can be great if you're on the road and coiling cable every day. When used as an install cable, flexible cables can bunch up. In addition, matte finish jackets that are very popular on these flexible cables can "grab" objects around them, which can be especially frustrating when installing cables in conduit.

It may seem counterintuitive, but for conduit installations, you really want cable that is fairly stiff and with a shiny jacket on the outside, so that it can slide easily down a conduit or over objects in its path. You can even *push* stiff, shiny cables down a conduit for a short distance, something you cannot do with matte jacket flexible cable.

Coiling Snake Cable and Reelers

Most snake cable used in the field is wound by hand into coils. The ideal way is to give a half twist to every other loop. This gives an even lay and prevents twisting of the cable. You can give that half-twist by pulling every other loop inside (instead of outside) the previous loop. One outside, one inside, one outside, one inside, and your cable will lay flat and be easy to uncoil for the next gig.

If you're on the road or have a remote truck, you might want to use hand-cranked or motorized reelers to reel-up and store cable between venues. These, of course, can't reel up cable outside-inside as suggested above. Some snake cable does not work well with these machines. Especially the motorized versions can sometimes put so much pressure on the cable innards that they begin have very weird lumps and bumps down the cable, creating a snake cable that is no longer round. These effects can happen without exceeding the pulling tension (see Table 3-13) of the cable.

This is an important concern with snake cables, and there is a considerable amount of analysis underway to determine why this happens. It is known that this effect is caused by the uneven tension on parts inside the cable. If one pair, or even one wire, is considerably shorter than the other components inside the cable, it will take much of the pressure from the reeler and force the parts around it to align, thus untwisting or trying to untwist, the other components.

Part of this problem is the way the jacketed pairs are "laid up" inside the cable. It is more of a problem with 16-pair, 24-pair, or larger pair-counts. In these cables, the pairs are laid in multiple layers in the cable. Snake cable seems to work better on a reeler when the layers are laid in reverse to each other, called *true concentric* layers, rather than layers in the same direction, called *mock concentric* or *unilay*.

This information is rarely found in a cable manufacturer's catalog or Web page, and some of the techniques to prevent this may be a trade secret. So you may have a hard time finding out from the manufacturer just how your snake was made. Of course, the ultimate, albeit expensive, test is to buy some snake cable and reel it back and forth.

Snake cables, even just analog snake cable, comes in dozens of choices and styles. Table 3-7 gives you just a few of the choices you have together with their key advantages and disadvantages.

Snakes and NEC Fire Ratings

As noted in Table 3-7, the NEC fire rating is one factor to consider. Recent changes in the NEC now require any installed cable to have a rating. Even cable now installed in a conduit must have a rating, even though there is no recommended rating.

This is a problem for many ultra-flexible snakes because most have no NEC rating, especially snakes made in Europe or Japan, which don't recognize the NEC, and therefore have no NEC rating. You can still use them outdoors or temporarily inside buildings, but, if the area you work in adheres to the NEC, all those unrated cables are now illegal to install.

TABLE 3-7 Installing Snake Cable

Style	Description	Advantages	Disadvantages
Ultra-flexible	Matte jacket, French Braid serve, or Reussen shield	Very easy to handle and hand coil	Very difficult to install, especially difficult in conduit; possibly no fire rating, so cannot be installed where NEC-rated cable is required
Flexible	Matte jacket, foil shield	Moderate handling, lower cost. Low fire rating (CM) if any	Difficult to install but allowed in conduit, if rated
Stiff	Shiny jacket, foil shield	Much easier to install, slides over equipment and slides through conduit, sometimes riser rated	Hard to coil; poor choice for trucks or on-the-road applications
Very stiff	Teflon innards, special fire-rated jackets	Meets plenum fire rating, no conduit needed if that area follows the NEC	Very expensive, very stiff, harder to install but not as hard as ultra-flexible

However, there are several riser-rated snake cables. This is only one step below plenum (see page 90–91). Riser-rated cables can go between floors without a conduit. These riser snakes are true install snakes, fairly stiff, with hard shiny jackets. They would be hard to handle on the road but are the ideal choice for installation.

Pair Counts

There is an on-going discussion about just how many pairs are the "right number." Of course, that depends on the application and the equipment chosen. There are snake cables available for 2 to 52 pairs and even higher. Table 3-8 shows some common pair counts and the applications for which they might be used.

TABLE 3-8   Snake Cable Pair Counts

Pairs	Applications	Considerations
2	Stereo-out, stereo microphones	Braid shield for microphones
4	Mixer inputs or outputs	If ins and outs go different directions
6	4-in, 2-out mixers	If ins and outs go to the same location
8	8-in mixers/recorders	ADATs, DA-88,
10	8-in, 2-out mixers	If ins and outs go to the same location
12	8-in, 4-out mixers	If ins and outs go to the same location
16	16-channel mixers/recorders	
24	24-channel mixers/recorders	Patch panels
26	24-in, 2-out mixers	Some patch panels
28	24-in, 4-out mixers	If ins and outs go to the same location

continued on next page

TABLE 3-8 Snake Cable Pair Counts (continued)

Pairs	Applications	Considerations
32	24-in, 8-out mixers	If ins and outs go to the same location
48	Large format consoles/recorders	
52	Large format consoles/recorders	

How to Buy Snake Cable

If you are doing an installation, there is one consideration you might not think about. Most of the time you will buy one or more snakes for each piece of equipment, a 16-pair for this old analog multitrack, an 8-pair for this little mixer, or a 24-pair for the patch panel. Each of these will be short, maybe even cut to length by the distributor. If you bought 8-pair snakes, you could do all of these things, two 8-pairs for the multitrack, one 8-pair for the mixer, and three 8-pairs for each patch panel. The advantages to doing this are many.

For one thing you would be buying a lot more of one kind of cable. This could lead to a serious cost reduction because 1,000 ft of one cable is cheaper than 100 ft of ten different cables. All the tools and associated parts, such as wire ties, labels, and heatshrink tubing, can be purchased for one size. Even multipin connectors can be standardized.

Sure, it's a bit more work to prepare three cables rather than one large cable for a patch panel, so you need to analyze the cost of installation (that is, labor) versus the cost of the cable.

Unbalanced or Balanced?

Many installations are forced to use a combination of unbalanced and balanced equipment. It is difficult to install, run, and maintain any system with two standards. Therefore, it is strongly suggested that you standardize on one system. The simplest way to decide is to count the inputs and outputs of the equipment you intend to use and determine how many unbalanced connections and how many balanced connections are in your system. Then you simply choose the one in the majority.

However, if you intend to expand or improve your system in the future, it is more than likely your future system will be based on the professional standard, that is, balanced lines. Therefore, if you can start now with a balanced system and integrate the unbalanced components into it, you will not have to make a major change later on. Also, this may encourage you to spend a few more dollars and buy equipment with balanced inputs and outputs.

Because many devices now run balanced lines on three-conductor phone plugs, you must read the manual for that piece of equipment to determine whether the inputs or outputs are indeed balanced. Because we're concentrating on wire and cable here, we won't go into the equipment choices except to say that some equipment pretends to have balanced lines because they use an XLR-style connector or a three-pin phone plug, and yet, internally, these have unbalanced wiring. Just be aware. If you want a professional quality end-product, you want a professional installation, and that means true balanced lines.

Balanced to Unbalanced and Vice Versa

Once you have decided which system to use, you may have to convert some of your gear from balanced to unbalanced, or vice versa. The boxes that do this conversion do not come cheap, especially if they are professional quality, so you may want to add their cost into your equation. The cost of this conversion often is greater than just buying the correct equipment in the first place.

There is one advantage to unbalanced systems. Because balanced systems are at a much higher level (Table 3-2), the devices that convert them to unbalanced are passive devices that require no power. Further, the quality of such a conversion can be very high, because no active components (ICs, etc.) are required. You can even build them yourself, if you have some technical expertise.

Converting from -10 dBv (consumer) to $+4$ or $+8$ dBu (professional) will be a bit harder. It requires active components because passive components cannot amplify a signal. This means a powered device, most commonly with RCA connectors for the inputs, and XLR or other balanced connection scheme for the output.

These devices are readily available from a number of manufacturers. It would be best to review their full specifications if you have any concerns about quality.

Impedance Matching and Conversion

We've mentioned that, at analog audio frequencies, the impedance of the cable is of little importance unless the cables are at least a mile long. However, that doesn't mean that impedance is a non-issue. Impedance between devices is very important. It is essential that the signal be transferred from the source device to the destination devices with the least loss.

A Match Made in Heaven?

How can you remember how to match? I used a little rhyme to help me remember which way the impedance needs to go:

> Low into high, and you'll get by
> High into low, just won't go

To the telephone company, impedance means matching a 600 Ω source to a 600-Ω load, but there is an even better combination for analog audio. To reduce the loading even further, most devices today have a very low output impedance. The input of the next device is a very high impedance, sometimes 10 kΩ or more. This means that very little load is placed on the source and all the energy can be transferred from one box to the next.

This combination greatly increases the effective distance of cables between devices, as shown in Table 3-1. As the source impedance gets larger or the destination impedance smaller, the load on the line will increase. Unfortunately, this affects high-frequency response. So high-frequency loss can be caused by a cable with too much capacitance, a cable that is too long, or improper loading of the output/input device.

Matching Transformers

If you absolutely must go from high to low impedance, you must use a matching transformer. There are many varieties available. Some can just be hard-wired inside or outside a box. Some are built with connectors on the box to simplify installation or allow quick connection. You also will get more level if you use a matching transformer when going from low to high impedance.

Probably the most common matching transformer has an XLR on one end and a phone plug on the other and is used to match low-impedance microphones to high-impedance mixers. Of course, this also converts from a balanced line (XLR) to an unbalanced line (phone plug), so you can kill two birds with one stone.

Balanced to Unbalanced

Any impedance-matching transformer will allow you to convert on either side to unbalanced. If you have three pins and a balanced line, just combine the ground pin with one of the other pins. That will be the ground connection. The pin left is the hot connection.

Because the two wires on a balanced line are identical, it doesn't matter which is ground and which is left as the hot pin. However, for consistency's sake, decide which is the hot pin in your installation and ground the other one. Of course, if you ground one pin at one end of a cable and ground the other pin at the other end, then both pins are grounded and you will get no signal at all. So decide on a pin!

There is an additional caution here. There are different kinds of balanced outputs. For instance, there are transformer-based outputs and there are active balanced outputs. The active balanced variety often does not respond correctly to the "short one pin to ground" method.

If you try the "short one pin" method and it produces odd results, try the second method. Just use the two wires from the balanced line. Use one as ground and the other as hot pin in the unbalanced connector. Do not attach either to the shield in the cable. Just let the shield 'float' unconnected.

If neither method works or performs oddly, it is strongly suggested that you contact the manufacturer of the source or the destination box, and ask for advice in unbalancing these connections.

Buy a Conversion Box

The most expensive but most predictable way to go from balanced to unbalanced or vice versa is a conversion box. This is a device that uses active circuitry going from −10 dBv unbalanced consumer to +4 dBm balanced professional. This often has a passive section to go in reverse. Because +4 dBm to −10 dBv is a decrease in level, amplification is not essential.

Pin 2 Hot

Most microphone companies have standardized pin 2 as the hot pin in a three-pin arrangement. Pin 1 is the ground, pin 2 and 3 are the balanced line, pin 2 is hot, and pin 3 is low.

ONLY IN SAN FRANCISCO

Just when you think this "which pin is hot?" argument is settled, there's some guy with a car in San Francisco and a license place that says "Pin 3 Hot."

If you wire your microphone cables identically, it won't matter that one wire is in pin 2, the other wire in pin 3, and pin 1 is the ground. However, if you have a cable that "flips" the pin, where pin 2 is hot at one end, but that same wire is connected to pin 3 at the other end, it can lead to some interesting and unpleasant consequences, such as phase cancellation.

Phase Cancellation

You've probably heard phase cancellation on the radio or TV and may not have realized what it was. It is most common in music or voice recordings originally recorded in stereo and played in monaural. Radio and television receivers, which have single speakers, automatically combine the left and right audio signals to create the monaural sound. You also can hear this effect on records, compact discs, tapes, and other formats.

What you hear is the middle disappearing. When someone is speaking or singing, that person's voice sounds very distant and the background sounds or musical instruments are much louder. The less prominent sounds like sound effects or solo musical instruments often are very loud. What is going on?

Here's what's happening. Most commonly, somewhere in the broadcast chain, or the recording chain if you hear this on a recording, there were two cables carrying the stereo information. One of those cables was wired up correctly at one end, but the other cable had its balanced line flipped at one end.

SECOND GUESSING THE PHONE COMPANY

Back in the dark ages, when I was a broadcast engineer, I would get lots of feeds from the phone company on analog lines. My biggest problem was, if I ordered a stereo pair, I could never predict how they would be phased. In stereo, it would sound just fine. So, when the remote started, I would be at the little monaural table radio at the receptionist's desk. If it sounded weird, I would flip a switch I had installed that flipped pins 2 and 3, but only on one channel, of course. I think the receptionist thought I was nuts.

In a stereo recording, the music or voice and other prominent content are placed in the "center." This means that the sound is at equal level on both left and right channels. But with one cable flipped, the two waveforms are not going up and down at the same time. When one is going up, the other is going down.

Here's the key; if you listen in stereo the sound seems pretty normal. (Some golden ears can hear phase problems

quite easily even in stereo.) But when combined, the left
and right channels no longer merge; rather, they cancel each
other! What you hear is the difference between channels, for
example, that odd instrument off to the side and that sound
effect off-screen right. Instead of left plus right, it's left
minus right.

This is why it is essential that you have some way of listening in monaural at some point to anything you make, or
any installation you work on.

Although the most prominent phase cancellation occurs
when listening to stereo, with one channel out of phase, on
a monaural box, it is possible to have one microphone out of
phase in an orchestral recording. When combined with all
the other microphones, it cancels out whatever is in common
around it. Good recording engineers can hear the effect. It's
almost like hitting the "solo" button. That player has a little
island of sound all around it that cancels out, instead of
adding to, the mix.

UTP and Analog Audio

There is one option available to designers and installers of
analog audio: unshielded twisted pairs (UTP). The phone
company has installed audio on UTP since the beginning of
radio and achieved some impressive results. A 15-kHz bandwidth with a signal-to-noise ratio of 60 dB is common performance. These analog services have been supplanted by
analog and digital microwave studio-transmitter links
(STLs) and T-1 lines.

The kind of UTP discussed here is intended to carry data,
commonly called *category* cables. The most common types of
category cable are Categories 3 and 5. Newer versions, such
as Category 5e and the emerging Category 6, are pushing performance levels of twisted pairs to quite amazing levels.

Category 3, with a bandwidth of 16 MHz, has been almost entirely relegated to telephone wiring, supplanting plain old telephone service wiring. FCC rules, as of July 2000, require that home wiring be a minimum of Category 3. A large percentage of telephone lines are now used for data at high-speed connections such as ISDN, DSL, and T-1. Concerns about signal emission from cables led the FCC to require the installation of Category 3. This has encouraged the design and installation of even higher-quality category cables in home installations. One of the fastest growing areas for Category 5 and beyond is high-end and custom homes. In addition, there is an interest in using such cables for more than just the telephone and the computer; for instance, running whole-house audio on Category 5. This is discussed in the next section.

Unbalanced Audio and Category Cables

We start with a comparison of unbalanced analog cables and Category cables. There is a huge range in the quality and price of analog interconnect cables, but they are not long enough for high-impedance unbalanced equipment, such as consumer analog audio gear. Table 3-1 listed a maximum distance of 27 ft, even with top-of-the-line coax. That might be fine for an in-room installation, but how about sending that audio to the master bedroom or the family room or duplicating what's playing in the home theater at other locations? The simple way is to reduce the impedance of the system and, while you're at it, go to a balanced line.

Don't Say *bay-lun*

Converting balanced to unbalanced requires the use of a device that converts from the balanced line of the UTP to the unbalanced line of a consumer interconnect. The device is called a *balun* (balanced to unbalanced). It is pronounced BAL-UN, not BAY-LUN.

Such baluns, which convert unbalanced analog audio to balanced Category 5 or higher, are becoming more common. My search found a number of manufacturers (see the Addendum); the most prolific is Energy Transformation Systems (ETS; 1-800-752-8208, www.etslan.com). They make baluns that convert one channel of audio up to four channels, the maximum on four-pair Category cables. One- and two-channel unbalanced versions are shown in Figure 3-4 as examples.

On one side is a male RCA or multiple RCAs (also called "phono" jacks). On the other side is an RJ-45 modular jack, the standard 8-pin connector for category cables. Category cables have up to four UTPs. As we will see, all of them can be used to carry audio.

Figure 3-4 Unbalanced audio baluns.
(Courtesy ETS, Inc. www.etslan.com.)

To determine the viability of running unbalanced analog audio down Category 5, or on the higher grade 5e or the emerging Category 6, one need only compare the requirements of such analog audio systems to the performance of these cables. Table 3-9 shows such a comparison.

TABLE 3-9   Category Cables and Unbalanced Audio

	Standard Cable	Category 5	Category 5e	Category 6
Format	Unbalanced	Balanced	Balanced	Balanced
Capacitance (pF/ft)	30	15	15	15
Impedance (Ω)	Not applicable	100	100	100
AWG	22/24	24	24	24
Shield	Yes	No	No	No
Measured crosstalk (dB) at 100 m (328 ft)	−90	−95	−100	Unmeasurable

The capacitance is listed as 30 pF/ft, but this is rather generous. Many of the standard interconnect cables, such as those included with receivers, amplifiers, and similar equipment, are often made with stranded aluminum center conductors and low-coverage spiral shields. They are often made entirely with PVC, inside and out, and have a capacitance of 50 pF/ft or even more. Of course, it is rare that these interconnects are longer than 2 m (6 1/2 ft).

Specially purchased interconnects can have much better quality, but at considerably higher price. Rarely are they better than the 15 pF/ft of category cables. Ironically, one of the key objections to the use of category cables is that they are *too cheap*, especially compared to the $100-per-meter high-end interconnects. Those with "golden ears" must be persuaded to listen to some Category 5 (or even better, Category 5e or 6) before making a judgment based solely on price.

The main problem is that the system is unbalanced, but the cable is balanced. So a balun is inserted to convert the source device to a 100 Ω balanced line.

Once converted, such an audio signal can go well beyond 1,000 ft with only minimal degradation (Table 3-1). The master bedroom or family room, or your friend in the middle of the next block, would be within easy reach with category cables. Of course, the impedance of the system is important but the impedance of the cable is not. It is the low capacitance and the balanced line that are providing the major advantages for Category cables at analog audio frequencies.

If the system to be converted is a balanced line, such as professional audio installations, the use of category cables becomes even easier. Table 3-10 compares standard analog twisted pairs and category cables.

TABLE 3-10   Category Cables and Balanced Audio

Format	Standard Cable Balanced	Category 5 Balanced	Category 5e Balanced	Category 6 Balanced
Capacitance (pF/ft)	30	15	15	15
Impedance (Ω)	Not applicable	100	100	100
AWG	22/24	24	24	24
Shield	Yes	No	No	No
Measured crosstalk (dB) at 100 m (328 ft)	−90	−95	−100	Unmeasurable

There is no requirement even for a balun. You can insert a category pair exactly where you would have put any audio pair. Inside an XLR, for instance, the balanced line is on pins 2 and 3. So you would connect the two wires from the category cable pair to pins 2 and 3. What about pin 1, the ground pin? With category cables, which are UTP, there is no ground, no shield, and no ground wire. So pin 1 is left empty. Rest assured, everything will work just fine. The balanced line is where the signal is. The shield was there to protect the balanced line.

And this, at least for analog audio, is a key point. The performance of a balanced line, whether older audio cable or newer category cables, is only as good as the balance at the source and destination equipment. If that equipment has poor CMRR, then you can make the world's best pair pick up noise or radiate its own signal into adjacent pairs. So check the CMRR of your equipment. Most modern gear has excellent CMRR, and many thousands of installations have been made running analog audio down UTP.

The real accomplishment is the dramatic improvement of the twisted pair. Instead of shielding an inferior pair with a foil and/or braid shield, we made a better performing pair.

How can an unshielded cable provide the noise rejection and crosstalk rejection of a shielded cable? The answer is shown in Figure 3-5. These actual laboratory test results indicate that analog audio on UTP can easily meet the crosstalk requirements of professional analog audio installations, to say nothing of the less stringent consumer/home requirements.

In Figures 3-5 and 3-6, the cable used in the test was a four-pair Category 5 patch cable, an excellent choice. Category 5 patch cable is stranded. Patch cable is the lowest-performance cable because it is stranded. In fact, the TIA/EIA 568-B standard for data cable uses its own set of requirements for patch cable, because it doesn't even come close to meeting the specifications of solid Category 5 cable.

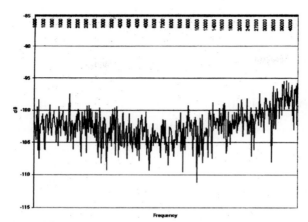

Figure 3-5 Audio frequency Category 5 FEXT.
(Courtesy of Belden Electronics Division, Inc. www.belden.com.)

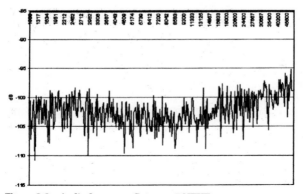

Figure 3-6 Audio frequency Category 5 NEXT.
(Courtesy of Belden Electronics Division, Inc. www.belden.com.)

For example, patch cords are allowed to have up to 20% worse attenuation than solid Category 5.

The only improvement with this particular cable as shown in these graphs is that the pairs are bonded. Bonded pairs improve noise and crosstalk since the distance between wires is one of the keys to CMRR.

Despite this choice of cable, the worst case of crosstalk is –95 dB at 48 kHz. In the 20-kHz range and below, it is even better at –97 dB and typically –100 dB. The trace shown in Figure 3-6 is the average of all possible crosstalk configurations (pair 1 to pair 2, pair 1 to pair 3, pair 1 to pair 4, etc.).

The laboratory measured far-end crosstalk (FEXT) of a 100-m cable (328 ft), where the signals are weakest (Figure 3-5). Is that where crosstalk will be most prominent? Or is crosstalk worst where the signals are strongest? That would be near-end crosstalk (NEXT), shown in Figure 3-6. Note that the NEXT numbers are virtually identical to the FEXT numbers. This is an average of all possible pair combinations.

You can run analog audio on category cables, and the higher the quality of the cable, the better the results. By the time you get to Category 6, the worst-case pair-to-pair crosstalk is unreadable even on a $60,000 Network Analyzer. The noise floor on this equipment is –110 dB, so the crosstalk is somewhere below that, perhaps –115 dB or –120 dB.

If you are using Category cables to run microphone level audio, be aware that this signal starts at –60 dB. Even if you use Category 6, with analog audio crosstalk below –110 dB, the signal-to-noise ratio at the microphone level is below –50 dB, which is good but not great. Of course, we don't know what the actual number is. How much lower than –50 dB is it? If it is –60 dB, that is acceptable. If it is –70 dB, that is very good. If it is –80 dB, that is excellent. Of course, the output level of the microphone might be –50 dB instead of –60 dB, and that would certainly help things.

This analysis of category cables will occur with every new signal we study. The true advantage is that you can use one kind of cable to do a myriad of tasks. Of course, this cable will run 10baseT or 100baseT computer data. That's what it was designed for. Category 5e or 6 cable is designed to even run the new 1000baseT (Gigabit Ethernet). Of course, it will run all the telephone signals, FAX, DSL, ISDN, and T-1 you might want. You can use it to run audio and even digital audio.

There's one other cable, flat cable, that we need to look at before we're done with analog audio cables.

Flat Cable and Analog Audio

Flat cable has been used for audio for many years. It is common to see it inside original equipment manufacturer (OEM) gear. The original flat cable was multiple conductors, with a jacket extruded over all the wires (Figure 3-7). The advantage to flat cable was the speed at which one could interconnect dozens or even hundreds of wires.

Most often, flat cable uses insulation displacement connectors (IDCs) which are configured to match the spacing of the conductors so that an entire ribbon of conductors can be connected at one time. Of course, there are specific numbers of pins in the connectors that correspond to the conductors of the flat cable.

Unfortunately, the crosstalk of multiple straight conductors is not very good. So most often they carry control signals, or DC, from switches or to lights, for instance. If these wires carry audio, it is for very short distances to minimize crosstalk.

There is one flat cable very similar to the multiple straight conductors shown in Figure 3-8, which has a shield layer under the cable. This layer reduces interactions

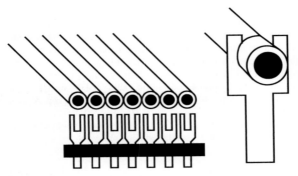

Figure 3-7   Standard flat cable.

between cables when they are layered, one over another. For higher-frequency application, this "ground plane" cable provides more stable impedance than regular flat cable. The conductor-to-conductor crosstalk is still poor, however. It also takes a bit longer to connect to ground plane cable, as

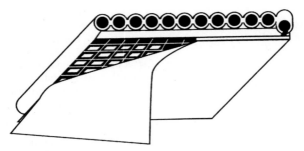

Figure 3-8   Ground-plane flat cable.

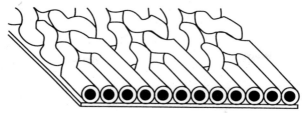

Figure 3-9   Twisted-pair flat cable.

the ground layer must be stripped away before the connector is inserted. There is a single wire, which lines up with the other wires of the ground plane cable, but is intended only to touch the shield layer and be the contact point.

There is a second type of flat cable, which consists of twisted pairs, as shown in Figure 3-9. These twisted pairs are then glued to a substrate, or plastic layer. Because you cannot crimp an IDC to pairs that are twisting, these cables untwist and go straight for a short distance to allow connections. Commonly, there are 18 inches of twisted pairs followed by 2 inches of straight and this pattern repeats along the entire cable.

Installing Twisted-Pair Flat Cable for Audio

All flat cable uses very small conductors. Flat cable has a number of common gages, the most common being AWG. These small wires and the resulting resistances, limit the distance that any flat cable can be run. And that's fine, because 28 AWG certainly is not rugged. This is no replacement for a regular snake cable. In fact, it should be used only in protected environments, such as inside equipment or, at most, between different pieces of equipment inside a rack.

As a rule of thumb, flat cable should run be no more than 10 ft. Table 3-11 shows the analog audio performance of the twisted-pair single-conductor version.

TABLE 3-11   Twisted-Pair Flat Cable Specifications for Audio

Parameter	Measurement Unshielded	Measurement Shielded	Comments
AWG	28	28	0.050-in spacing
Capacitance (pF/ft) at 1 kHz	16.7	16.8	
Crosstalk (dB) at 20 kHz	−80	−70	10-ft length

Whereas impedance is not important at analog frequencies, it is important for digital audio, which we will discuss in Chapter 4.

This twisted-pair flat cable has 18 inches of twisted pairs followed by 2 inches of straight conductors. This means that the minimum distance is 18 inches between devices, but you can use any multiple of 18 inches after that.

Table 3-12 lists the basic analog audio cable types for home theater and professional installations.

The maximum pulling tension can be calculated by knowing the number of wires in the cable and their gage sizes. Table 3-13 shows the recommended pulling tensions.

The information in Table 3-13 is fairly accurate for multi-pair snake cable and similar single conductor and paired cables. It is less accurate for coax cables, especially where braid shields have many small conductors. They have a lay length (twist length) that is much looser than the stranded or solid conductor in the center. Therefore, it is probably safer with the center conductor as the only pulling element in these calculations. Further, very small cables require very

TABLE 3-12 Installing Analog Cable

Type of Cable	AWG Range	Cable type	Application	Key Parameter	Key Parameter for Installation
Microphone	28–20	Twisted pair, braid shield	Mic wiring	Capacitance	Not installed
Line or mic level	26–22	Twisted pair, foil shield	Interconnect	Capacitance	Size
Line or mic level	26–22	Twisted pair, foil shield	Snake cable	Capacitance	Less flexibility is better
Line or mic level	24	UTP, Category 6 if mic level	Interconnect	Capacitance	Small, fast
Home theater	Any	Coax	Interconnect	Capacitance	Unbalanced
Home theater	24	UTP	Interconnect	Capacitance	Requires balun

light tension when being pulled. There is more information on pulling strength and how to set up a pull in Chapter 9.

TABLE 3-13   Pulling Tension

AWG	Maximum Pulling Tension (lbs)
24	4
22	7
20	12
18	19
16	30
14	48
12	77

Splits and Mults

Splits and mults refer to "splitting" one output or having "multiple" feeds from one source to two or more other devices. Analog audio cable is very forgiving in terms of splits and mults, where you have more than one pair attached to a single output terminal. Attaching two or more pairs allows the same audio signal to be fed to any number of devices. There are only a few negative points.

Attaching multiple cables affects the impedance of the cable. That doesn't matter because impedance of the cable is not important at audio frequencies. What does matter is whether the source and destination impedances are close to each other because the level on each pair will change, as shown in Table 3-14.

These losses are not dramatic and can be easily compensated for. Further, if the output impedance of the source is relatively low, less than 100 Ω and the input impedance of each destination device is more than 10 kΩ, then there is

almost no loading on the source and splits can be made without any appreciable loss.

TABLE 3-14 Split Levels

Number of Outputs on One Terminal	Level on Each Pair versus a Single Output
1	0 dB loss
2	−3 dB loss on each
3	−4.5 dB loss on each
4	−6 dB loss on each

As we will see, where impedance is important, such as
digital audio or analog or digital video, splitting or multiple
outputs is a much more complex problem.

Even with low-source impedance, a short circuit on one
pair in a multiple-output configuration is the same as a
short on all pairs. That is, the failure of one pair in this way
would render all outputs inoperative. This is especially bad
when the multiple outputs feed different studios in a multi-
studio environment, such as a broadcast or post-production
installation. Then the short on one pair may be a long distance away, possibly even on another floor. You would have
to separate the pairs from the output and look at them with
an ohmmeter just to tell which is the offending pair.

Manufacturers make specialized transformers that can iso-
late the outputs. Then, if there is a short or open on one pair,
the other pairs will not be affected. But these are passive
devices and cannot make up for the loss of each split, which is
increased by the natural loss within the transformer.

It is for this reason that most installations use distribution amplifiers. These are devices that take one, or sometimes more than one, input and distribute it through multiple outputs, with each output fed by an active amplifier

channel. Thus, if one output pair fails, whether the failure be a short, an open, or an unbalancing (one side shorted to ground), it has an effect only on that amplifier and no effect on any other output.

Distribution amplifiers also can be used to make up for the level loss of splitting or multing, so that all outputs are at whatever level you desire. If you have multiple studios, sending different levels to each studio might be helpful. Some mixers or consoles may require +4 dBu levels, others may be set up for +8 dBm. Still others may use equipment intended to run at −10-dBv consumer levels. Distribution amplifiers can deliver all of these levels simultaneously.

Connectors

Since the impedance of audio cable is of little importance, the impedance of connectors for audio cable is even less important. This is the reason the impedance of any popular connector is unknown and, as far as I know, has never even been measured.

The main parameters for balanced connectors are listed in Table 3-15.

TABLE 3-15   Analog Audio Connectors

Connector	Ruggedness	Assembly	Strain Relief	Weatherproof?
XLR	Excellent	Good	Poor–excellent	Available
TRS	Very good	Poor	Good–excellent	No
Bantam	Good	Very poor	Good–excellent	No
Stereo phone	Good	Good	Poor	No
DIN	Fair	Poor	Poor	No
Sub-D	Good	Poor	Fair	No
Elco/EDAC	Very good	Good	Good	No

Of course, if you have a box that requires a specific connector, then that is what you will use, whether you like it or not. And, within each of the connector types listed in Table 3-15, there often can be a very wide range of quality and price.

Table 3-16 shows the most popular unbalanced connectors.

TABLE 3-16   Analog Connector Properties

Connector	Ruggedness	Assembly	Strain Relief	Weatherproof?
RCA	Poor–excellent	Poor–good	Poor–excellent	No
Mono phone	Good	Fair	Poor–excellent	No
Mini-phone	Fair	Poor	Poor–good	No
BNC	Excellent	Good	Good–excellent	Available

Perhaps an argument can be made that consumer home theater installations need less ruggedness. However, the quality of much home equipment is approaching, even equaling to, that of professional pieces. In many cases, the balanced–unbalanced difference is the only discriminator between professional and consumer systems.

It is no wonder that many connectors developed for professional cables are used in home installations. This is especially true of precision video cables, where there are a number of very high quality crimp and clamp-style RCA connectors. Some of these are functionally identical to BNCs intended for the same cable, but with RCA dimensions up front.

As with cable, there are RCA and other unbalanced connectors selling for astronomical prices. If you can bear the difference, buy them. I would suggest that high-end installers find out what BNCs cost for a particular cable and expect the RCA version to be in the same range.

75Ω RCA

There are more than a few manufacturers who claim to have 75Ω RCA connectors. Although the cable attached to them might be 75 Ω and the section of the connector attached to the cable might be 75 Ω, the working end of the RCA is definitely not and can never be 75 Ω.

That's just fine because, for analog audio, the impedance doesn't matter. It's just a connection. Check out ruggedness, ease of insertion and removal, serious gold plating on the center pin (>30 microinches), and ease of connection to cable. Those are the important features for analog audio.

Speaker Cables

No subject about consumer audio is more confused and convoluted than speaker cables. This is especially true in the high-end and home theater worlds. Speaker cable can routinely go for $100 or more and feature exotic metals or sometimes no metals, exotic plastics, patents, trade secrets, and lots of pseudoscience thrown in for good measure.

Let's start with what we know and what is important in connecting speakers.

Amplifier output

Amplifier outputs are generally marked with impedance values such as 4, 8, or 16 Ω. In fact, the output impedance of any decent amplifier is many times smaller than these numbers, certainly a fraction of an ohm. This is desirable as it delivers all the power to the speaker.

The output of an amplifier, fed to a speaker that actually has one of those three output impedances, can be substantial. Table 3-17 shows the current (in amps) for various wattages.

TABLE 3-17   Amplifier Wattage and Current Output

Output Impedance	4-Ω Load	8-Ω Load	16-Ω Load
1 watt	0.5 amp	0.35 amp	0.25 amp
10 watts	1.58 amps	1.12 amps	0.79 amps
100 watts	5 amps	3.53 amps	2.5 amps

NOISE ON A SPEAKER CABLE

There was one time when I heard noise on a speaker cable. The system ran some column speakers with small wire (20 AWG). It ran the cable right past a bank of SCR dimmers. This was back in the 1970s, when SCR dimmers were new. They actually fed enough RF into the cable that you could hear it in the speaker. Of course, just moving the cable away ("inverse square law" solved the problem. Shielding? These cables were shielded mic cables so you could use them to run mics or the speakers. A bad idea all around. The shielding did nothing when used as speaker cable.

Table 3-17 clearly shows that the signal running down a speaker cable, as opposed to a microphone cable or line-level interconnect, is huge. Therefore, it is difficult to interfere with this signal—not impossible, just difficult. Of course, if you were going to invest money into cable to reduce noise, the best place would be where the weakest signal is, that is, the microphone cable. Investing in shielded speaker cable in most cases is wasted time and money. For one thing, it would take a huge amount of noise to affect a speaker signal. Such a huge signal would overwhelm a braid, foil, or combination shield, which is the reason unshielded speaker cable is the norm.

Shielded or twisted-pair speaker cable

Many installers think that twisted-pair speaker cable is better because it rejects noise. Now that you know about balanced lines (page 73) you know that balanced lines work only where there are identical signals of opposite polarity on the two wires or, more precisely, where the impedance on each wire (and attached equipment) is at the same impedance relative to ground.

This is not the case with speaker cable. One terminal is hot (positive), and the other is at zero potential (ground). In a balanced line the negative value would be equal to the positive value. This is the case with speakers.

DAMPING FACTOR

There is one other reason to use large-gage cable for running speakers, and that involves the transfer of power from the amplifier to the speaker. There is a specification called the *damping factor*, which is the speaker impedance divided by the output impedance of the amplifier. Speaker cable with lots of resistance, such as very small gage or very long runs, add to the speaker impedance and can affect the damping factor. Damping factor starts as a large number (~1,000) and drops as frequencies get higher. At high frequencies the resistance of the speaker cable can add seriously to the damping factor. This also affects the *slew rate*, which is the ability of the amplifier to deliver very fast risetime (high frequency) signals. Therefore, use bigger wire.

If you built an amplifier with a balanced line output, which would be easy to do, you would have to have a balanced input speaker to feed it to. This also would be fairly easy to do. What would you gain in noise reduction? Probably nothing, because the speaker signal is already so large.

Twisted-pair cables, especially twisted pairs with a jacket, are more rugged and easier to pull and install but offer nothing in terms of added performance in speaker applications. To be sure, there is only one agreed-to parameter with speaker cable: gage size.

Bigger is better—this makes sense when you are trying to deliver amps of current. You don't want the signal from the amplifier used up in the speaker cable, so bigger is better. How much bigger? The most popular speaker cable size in high-end installations is 10 AWG. In commercial installations where speakers are wired directly, the most popular size is 12 AWG.

Speaker cable loss

Table 3-18 below shows the percentages of amplifier wattage lost in a speaker cable. In this example, the speaker is wired with 20 ft of stranded cable, which means 20 ft to the positive terminal and 20 ft back to ground.

TABLE 3-18   Speaker Cable Loss

Load	20	18	16	14	12	10
			AWG			
4 Ω	9.8%	6.5%	4.2%	2.7%	1.7%	1.1%
8 Ω	5.2%	3.3%	2.1%	1.3%	0.9%	0.5%
16 Ω	2.6%	1.7%	1.1%	0.7%	0.4%	0.3%

Of course, if 0.5 percent is too much to lose, you can always use even larger wire. You can also reduce the resistance by moving the amplifier closer to the speaker, which is one of the advantages of the numerous powered speakers now available.

One can also determine the amount of loss you are willing to accept and then calculate the distance in feet for each gage cable. Table 3-19 shows these footages.

TABLE 3-19   Speaker Cable Distance

Cable AWG	4-Ω Load			8-Ω Load		
	11% (0.5 dB)	21% (1 dB)	50% (3 dB)	11% (0.5 dB)	21% (1 dB)	50% (3 dB)
12	140	305	1150	285	610	2,285
14	90	195	740	185	395	1,480
16	60	125	470	115	250	935
18	40	90	340	85	190	685
20	25	50	195	50	105	390
22	15	35	135	35	70	275
24	10	25	85	20	45	170

Courtesy of Belden Electronics Division www.beldon.com.

Parallel speakers

Many amplifiers have an A+B switch to power multiple sets of speakers. You can also hardwire speakers in parallel, thus powering more than one speaker from one output of an amplifier. There are only three cautionary notes to such wiring.

First, with more than two speakers running from one output, anything which happens in one speaker will affect the other. If there is a short circuit in one speaker, both will stop working. If there is an open circuit in one speaker, the volume of the other may change. If there is an intermittent open circuit, the other speaker cable will change in volume, going up with an open circuit and down with a closed circuit.

However, if the actual output impedance of the amplifier is very low, the changing load of an intermittent speaker might not be heard. It is interesting to consider that an amplifier could be built so well that you wouldn't hear a speaker fail!

The second consideration is the fact that, by paralleling more than one speaker, the effective impedance will drop by a factor of 2, and the current drawn from the amplifier will go up by a factor of 2. Table 3-20 shows the effective impedances of multiple speakers wired in parallel.

TABLE 3-20   Parallel Speaker Impedance

Number of Speakers	4-Ω Impedance (Ω)	8-Ω Impedance (Ω)	16-Ω Impedance (Ω)
1	4	8	16
2	2	4	8
3	1.33	2.66	5.33
4	1	2	4
5	0.8	1.6	3.2
6	0.67	1.33	2.67
7	0.57	1.1	2.28
8	0.5	1	2

Most amplifier manufacturers will advise you not to run amplifiers with less than a 2-Ω load. As you can see, that is easily accomplished with low-impedance components, such as 4-Ω speakers. Only the 16-Ω load allows you to add a few speakers in parallel before you get to that critical value, and 16-Ω speakers are fairly hard to find.

In truth, the actual impedance of a speaker is not constant and varies throughout the range of frequencies that are reproduced. Some speakers, by themselves, fall to the 2-Ω limit or sometimes lower. Unless you obtain a graph of the speaker impedance versus frequency from the manufacturer, you have no way of knowing just what your speaker will do. If you parallel that speaker with other speakers, you may find out in a dramatic way that you have a problem.

Most likely, your amplifier has some type of protection circuit, circuit breaker, or fuse. If your installation falls to some impossibly low impedance, it will look like a short circuit to the amplifier, and the protection device will be triggered.

If you are playing music, the real clue is that the protection will start at a specific note in the music. That is the frequency where the impedance is the lowest and the amperage is the highest. Those protection devices are current-driven and will work when that current limit is exceeded.

In conclusion, avoid running speakers in parallel. Run as few as possible when it can't be avoided. You also might consider changing to a distributed loudspeaker system, which we will discuss in the next section.

Exotic compounds and constructions

There are many unusual and exotic versions of speaker cables. These include mixed gages of wire, high-purity copper, and combinations of metals (silver, copper). These can affect the overall resistance of the wire, but there is no correlation between how these cables "sound" and how they measure in the laboratory. In fact, there is a patent issued on mixed-gage speaker cables.

Many readers would think that it is necessary to listen to speaker cables to make a selection. Of course, as a consumer, you are free to do whatever you wish. Just be aware that you are now entering an area where psychology affects your selection and where measurements and science do not.

A GREAT EXPERIMENT

I have always wanted to run an experiment regarding speaker cables with mixed gages. We are told by these manufacturers that the large wires are for low frequencies and the small wires are for high frequencies. So, strip both wires at one end and connect an AC plug as you would on an extension cord. At the other end, separate the small wires (high frequency) from the large wires (low frequency). Now ask the hi-fi salesperson to hold onto the small wires as you insert the AC plug into the wall. Of course, power is at 60 Hz, a very low frequency, so it won't be on those small wires. For some reason I have yet to find a salesperson who is willing to hold on to the high-frequency wires. I wonder why?

Once you have selected all other components, especially speakers, you can easily make A/B comparisons. Making blind comparisons, that is, not knowing which cable is which, is always preferable.

Anyone who buys anything because it is more expensive than a similar piece or rejects something because it is less expensive is not making a rational decision. The law of diminishing returns certainly applies here because smaller and subtler changes have higher and higher price tags.

Most would agree that investing the same money in better speakers would be more productive. In any listening environment, repositioning the speakers in relation to the listener will have a dramatic result, more so than the choice of speaker cable.

Consider this: Just where did this music come from? How did it get to you? What kind of wire and cable did it go through to get there? Too often, listeners think that, choosing the original microphone cable finishes the job. Professionals in the recording and duplications industries will see the humor in this. There are many steps between the original microphone and your living room. Even a live broadcast on radio, television, or the Internet has dozens of different cables, including no cables (broadcasting), to pass through. It's unlikely that all these cables were the $100-per-foot variety. So what makes the difference?

Constant voltage systems

In large installations with many speakers, you have an option called *constant voltage* or *distributed* loudspeaker systems. The reason these exist and the benefits they offer are immediately apparent.

As demonstrated with amplifiers running 4-, 8- or 16-Ω speakers, the key is cable resistance. Amplifier output impedances are so low, much lower than 4, 8 or 16 Ω, that even large wire has a significant effect on amplifier and speaker performance. The solution is to change the impedance of the system.

This can be easily done with transformers. They can be added outside of the amplifier, or you can buy an amplifier with them built-in. These are specified by the "voltage" of the system, which makes sense because we are also converting from a current-delivering system to a voltage-delivering system.

The most common version is the 70-volt line, but there are 25-volt and even 100-volt systems. We will concentrate on the 70-volt systems but be aware that those other voltages, while harder to find, can be obtained. Each has advantages and disadvantages.

However, there is one insurmountable disadvantage to these distributed systems: added cost. This cost is not just the extra transformer inside the power amplifier but also the fact that each and every speaker to be fed will also require a transformer to go from 70 volts back to 4, 8 or 16 Ω.

Moreover, the quality of reproduction in such a system will be determined in large measure by the quality of these transformers. It is no wonder that the majority of installations using these systems only use them for marginal-quality applications such as background music or paging.

However, high-quality 70-volt transformers do exist and are often used where long runs of high-quality audio are required such as in theater, auditorium, or stadium installations.

The key advantages to distributed systems is distance. Table 3-21 shows the distance in feet versus gage size. Compare these distances to those in Table 3-19 and you will see the difference.

In Table 3-21, the calculations were based on a 25-watt 70-volt system. The effective impedance of such a system is 196 Ω. What is immediately apparent is that, compared to 196 Ω, the resistance of the cable is pretty close to inconsequential. Even the smallest wire listed, 24 AWG, at the lowest loss (0.5 dB), can go 520 ft. This means that you could wire up these speakers with telephone wire, and they would work just fine.

TABLE 3-21 70-Volt Distance

Loss	11% (0.5 dB)	21% (1 dB)	50% (3 dB)
12 AWG	6,920	14,890	56,000
14 AWG	4,490	9,650	36,300
16 AWG	2,840	6,100	22,950
18 AWG	2,070	4,450	16,720
20 AWG	1,170	2,520	9,500
22 AWG	820	1,770	6,650
24 AWG	520	1,120	4,210

70 VOLTS ON A BUDGET

I once had to install a background music system with almost no budget. I got the speakers from a church that was throwing them out. I bought some really cheap 5-watt 70-volt transformers from a catalog house. The only real money I spent was on two 50-watt 70-volt transformers. Because I didn't want to spend money on an amplifier with a 70-volt output built in, I simply took an old hi-fi integrated amp and fed each output into one of the big transformers, turned around. That way, the 8-Ω output of each channel fed the 8-Ω side of the transformer. The other side was 70 volts and away I went.

Wattage taps and parallel speakers

The problems that arise from running 8Ω speakers in parallel disappear with 70-volt systems. Ohm's Law (see Addendum) says that voltages in parallel are the same. Therefore, you can run as many speakers in parallel as you wish, and they will be delivered the 70-volt signal they need. The only limitation, of course, is the total wattage available from the amplifier.

This is the reason each 70-volt transformer is rated by wattage. You choose the wattage appropriate for each speaker. In general, these transformers come in 5-watt, 10-watt, and 20-watt versions, although larger wattages are available. But that's just the beginning, because each transformer has "taps" on it, connections for even less wattage. For instance, a 5-watt 70-volt transformer might have taps on it such as "0.5 watts, 1 watt, 2.5 watts, 5 watts," each with a terminal, and a common or ground terminal. You will hook up your wires to the common (ground) terminal and one of the taps. But which should you use?

---

70 Volts and Impedance

Remember that rhyme "Low into high and you'll get by"? Well, that also applies here. You can actually run an 8Ω amplifier output directly into a constant voltage system as long as you have enough power. A 100-watt amplifier (at 8Ω) matches a 25-volt line, and a 500-watt (8Ω) amplifier matches a 70-volt line. It is still not as efficient as having a dedicated 70-volt output built into an amplifier, but it will work.

The ability to choose taps is a real bonus in distributed systems. This means you can choose how loud this particular speaker is compared to any other speaker. If you have a dozen speakers hooked up to the same output of an amplifier, each speaker can have its relative volume chosen by selecting a tap.

If you have an area that requires a low level, such as a reception area, and one that needs to be loud, such as a bathroom, you can accommodate both simply by choosing taps. Further, you can go back at any time and adjust the levels by moving the tap.

This is a good reason not to solder the taps in place, although you might be tempted from a reliability standpoint. Most 70-volt transformer taps will take push-on terminals, such as those used in car fuse boxes. Then you can simply pull the tap from one setting and move it to another or another.

The trick is to set the overall level with the master volume control of the power amplifier and then adjust the zones by moving taps on the transformers. Just to be safe, make sure that the total wattage of all the transformers does not exceed the total wattage of the power amplifier.

If you buy background music speakers, you can recognize them by the small metal flat sections, with two holes, just appropriate for your 70-volt transformer.

Don't forget plenum!

Don't forget that you're probably installing these speakers in a drop ceiling. If this is a commercial installation, that drop ceiling is a "plenum" area. If your community subscribes to the NEC, then you must use plenum-rated cable to wire up your 70-volt system. But that's not a problem because plenum-rated pairs are available in almost any size. In addition, because even small wire works just fine for 70-volt systems, you will save a considerable amount compared to big wire. You can easily buy plenum-rated telephone cable and use that to wire 70 volts *and* the telephone.

So what about 25-volt and 100-volt systems?

There are other less common distributed systems based on other voltages. The 25-volt version provides even cheaper and smaller components, but the distances are considerably shorter than in the 70-volt (Table 3-15). The key to a 25-volt system is to stay under 30 volts, which is a significant dividing line in cable voltage ratings.

The 100-volt system is just the opposite. It will go considerably farther than the 70-volt system. In large stadiums and similar venues, the added distance may be a real help. The transformers are much larger. Because of the higher voltage, this system most commonly uses wire and wiring practices intended for electrical power connections.

Digital Audio

What Is Digital Audio?

If I had a magic wand, I would remove the word *audio* from "digital audio." We have had more than 100 years to talk about analog audio, and many of the de facto rules of audio have been ingrained within our brains to the point that they have become immutable and unbreakable "law."

In truth, digital audio is more "digital" than "audio." Any designer or installer who thinks that digital audio is just another form of audio is in for a rude awakening, if not some serious or fatal flaws in any digital audio installation.

ONES AND ZEROS

OK, OK, now don't you digital geeks write me a note. The bit stream for digital audio is not really a stream of ones and zeros. It uses a system called NRZ, non-return to zero, that means it's really a stream of ones. It only returns to zero at the transitions. This is a simple way of compressing the data and speeding up delivery. But it puts more weight on the clock, because the transition points have to be accurate to the bit. If you're off by one, every bit beyond in that section will also be off.

Digital audio is a bit stream, a series of ones and zeros, no different in these basics than a bit stream from a word processor, a spreadsheet, or the data your bank has on your account. They are all data. The fact that this particular data is audio is of no concern for the wire and cable or for many other passive components. That's not to say that digital audio cable is not different from any other cable. Of course, it is manufactured to a standard, but you could easily design a system that uses the same cable for a completely unrelated data application and it would work just fine.

So part of what we will be doing in Chapter 4 is unlearning many of the rules we learned in Chapter 3. As Bette Davis said, in *All About Eve* "Fasten your seatbelts. It's going to be a bumpy night!"

Why the Rush to Digital?

It seems that everything is going digital. Eventually even radio and television will be all-digital. Many recording studios are already all-digital. Hollywood is rapidly moving to all-digital. So what happened and why?

The cost of going digital has been coming down dramatically. Digital solves one major problem: multiple reproductions. Perhaps you've tried to copy a CD onto a cassette. You might have noticed that the cassette never quite sounded as good as the CD. Then, if you dubbed your cassette for a friend, it sounded even worse. By the fourth or fifth dubbing, that tape was terrible and unlistenable… unless it was digital. Digital transmission is more robust than analog. Digital signals can be regenerated with very high accuracy. In addition, error correction allows an imperfect signal to be functionally perfect.

As long as the ones and zeros of the digital signal are transferred, stored, and reproduced accurately, the 5th or

105th copy will sound exactly the same. It's just a question of preserving the bit stream. Theoretically, a signal could go through 100 devices, be converted to a dozen different coding schemes or modulation types, and go up and down from a satellite, on fiber optic cables underground, or on copper cables in between. As long as the bits make it, and none are lost or introduced in error, the resulting signal will sound exactly like the original source material.

Data Rates and Standards

Digital audio began in the 1960s with work by the Bell System. People such as Harry Nyquist and Claude Shannon of Bell Laboratories developed most of the rules surrounding digitizing audio such as sampling and aliasing. They were trying to develop multiplex phone calls, that is, put many phone calls on a single cable. While this could be done in the analog mode, by shifting each call to different frequencies, work with digital signals and computers made it apparent that the ideal way was to convert analog signals to digital. In the digital mode, each call is a bit stream, a series of ones and zeros, and could easily be mixed with other calls.

The resultant DS-1 cable, a shielded twisted pair, can carry 24 digitized voice channels (DS-0). The total digital bandwidth is 1.544 megabits per second (Mbps; sometimes written Mb/s). Not content with that, they later added a DS-3 channel, interconnected with microwave, fiber, or a specialized 75Ω coax. This carries 28 DS-1 channels (672 voice channels), with a total data rate of 44.736 Mbps.

Along the way, it became apparent that the quality of the received signal was directly related to how the original analog signal was converted to digital. The conversion process "samples" the analog signal, or slices it into parts. These samples are then converted to digital "words." The rate of the slicing, or sampling, is critical.

The reason is simple; our ears are analog. So the system, even a digital system, at some point must convert back to analog so we can listen to the sound. Likewise, every microphone and every speaker are analog. Even the latest digital microphones start with a diaphragm moving to the compression and expansion of the air by sound waves. After that, you can sample that analog signal and convert it to digital.

All audio signals must start and end as analog signals because we live in an analog world. One day, we may have digital implants connected directly to our brains, but I would bet that our nerve impulses are analog, since our brains are products of billions of years of evolution in an analog universe. Even those implanted circuits will have to convert back to analog for the final connection.

If sampling rates, the rates at which the analog signal are converted, are critical, how can we decide what rate to use? There is no answer to this question, and that is why there is no end to the debate on the appropriate sampling rate. It is also the reason the standard has been expanded to include new data rates and then expanded again.

In the 1970s digital recording was done with specialized one-of-a-kind custom systems, and the standard for digital audio was not ratified by the Audio Engineering Society (AES) and the European Broadcast Union (EBU) until the early 1980s. This international standard, commonly known as AES/EBU, set down the original requirements for recording and playback of digital audio.

Table 4-1 shows a list of the most common sampling rates. The sampling rate is expressed in hertz because it is determined by a "clock," an electronic frequency generating circuit. It is the control for the sampling.

The common sampling rate in the 1970s was 250 kHz. Although this produced astounding results, it used up a lot of tape. The question of how much tape to use is still important when deciding on a sampling rate. These days, how much hard-disc space to use may be more pertinent.

TABLE 4-1   Digital Sampling Rates and Applications

Sampling Rate	Common Application
32 kHz	Voice recordings, reportage
38 kHz	Music, FM station quality
44.1 kHz	CD sampling rate for music
48 kHz	Audio tracks on professional video machines
88.2 kHz	Double the quality of 44.1, easily generated
96 kHz	Top-of-the-line professional machines
192 kHz	Double quality of 96, emerging technology

The only other question is, What quality of reproduction do you wish to achieve at the end of the process? The sampling rate chosen is directly related to the quality of the final product.

That isn't to say there are no other factors to consider. Chips that convert from digital to analog or from analog to digital (commonly called D/A or A/D converters, respectively) are getting better and better. With each generation they do a better job, with less distortion, and fewer self-generated artifacts. But it doesn't alter the fact that sound recorded at a 32-kHz sampling rate will never be the same as sound recorded at 44.1 kHz, much less 96 kHz.

The reason these rates sound different is really the center of the battle between analog and digital. With analog systems, there is essentially no limit to the frequencies that are recorded. More accurately, the limit is set by the design considerations of the equipment. You could easily build a tape recorder that could record up to 40 kHz, and you could build a system that could reproduce that wide a bandwidth.

With digital there is a hard and fast limit to bandwidth. It's called the *Nyquist limit*. (Remember Harry Nyquist from

Bell Labs?) It says that the highest frequency that can be recorded and reproduced is one-half the sampling rate. Table 4-2 shows the same list of sampling rates as Table 4-1, but with the maximum frequencies that could be reproduced from the speaker at the end of the chain.

TABLE 4-2   Digital Sampling Rates and Bandwidth

Sampling Rate	Maximum Reproduced Bandwidth
32 kHz	16 kHz
38 kHz	19 kHz
44.1 kHz	22.05 kHz
48 kHz	24 kHz
88.2 kHz	44.1 kHz
96 kHz	48 kHz
192 kHz	96 kHz

NOTE ON A TELEPHONE POLE

In the mid-1980s, when CDs and DAT tapes really started to come into prominence, someone tacked a manifesto to a telephone pole near where I lived. It warned everyone that the then-new digital audio systems filtered out all those "special" frequencies that you couldn't hear. It was a government conspiracy! It was a worldwide conspiracy! You could process an record, or even an Edison cylinder, and get all those hidden frequencies, but with digital, it was a "brick wall." Nothing after that Nyquist filter. We were supposed to write our Congressman and demand a ban on digital. Oh, well.

Aliasing

Why can't we go past one-half the clock rate? Because frequencies sampled beyond it can produce aliasing. Using an alias instead of your real name is a lie, a deception. In audio aliasing means there are frequencies that are not supposed to be there, frequencies "manufactured" by the circuit itself, and not present in the original audio content.

Aliasing, especially when it is bad, sounds so alien that it is instantly recognizable. You can hear it on long-distance phone conversations, or when digital cell-phone circuits have problems. The other person sounds like Donald Duck, but at dozens of different tones simultaneously, like the Borg talking on *Star Trek*. To avoid this, an audio filter, sometimes called a "brick wall" filter, is placed in the device to remove all the frequencies after that one-half-the-clock frequency.

We're limited to a maximum audible frequency of one-half the clock, so we're back to the analog audio argument. What can people hear? Because each of us is different, especially as we get older, our ability to hear high frequencies changes. I am sure there are people out there who can hear beyond 20 kHz. And perhaps there is truth to the argument that you need the harmonics beyond hearing to complete the full-spectrum experience. The point is, you have a whole lot of different bandwidths (and therefore sampling rates) to choose from. As the designer or installer of a digital audio system, the sampling rate might be a choice you will have to make!

The Digital Cliff

This is our first visit to the digital cliff. The digital cliff is not a place, it's an effect that occurs on cable. It applies to all digital signals. Although discussing it in relation to digital audio, where it is an important effect, it is a critical effect for digital video and high-definition television.

One of the main problems with digital signals and systems is that it is very hard to tell how well everything is working and how close to not working everything might be. For instance, as the cable gets longer, and the level drops (and other effects become more prominent), bit errors will increase on the cable. But you won't see this because digital signals are an all-or-nothing proposition. As long as the receiving device can tolerate, or correct, these errors, the result will be a reasonably perfect signal. But there comes a time when the cable is just too long, or the wrong impedance, or high capacitance, and bit errors are generated that cannot be tolerated, much less corrected. At that point the system will shut down.

Just imagine you have a cable run for digital audio that isn't working. You might think the connectors are bad or were put on incorrectly. Maybe the source or destination equipment was set up incorrectly. You know, one of those dip switches inside was set wrong. Or maybe there was something wrong with the source material, the signal itself.

What if the cable is simply too long, and how could you tell? It would take a lot of testing, and a bit of frustration, before you realize that you have exceeded the maximum distance. The distance from a "perfect" signal to shut-down might be as short as 10 ft.

Further on we show charts for twisted-pair and coax cables running digital audio signals. These distances are maximum distances designed to keep you away from the digital cliff. The actual digital cliff is approximately double these distances.

If you need to go farther than recommended, simply buy a piece of test gear that will generate an AES signal and show the bit errors at the other end. If you see no bit errors or, more precisely, an acceptable bit error rate, then you're fine. Table 4-3 shows bit error rates and how often an error might be produced.

TABLE 4-3   Bit Errors and Sampling Rates

Bit Errors	44.1-kHz Sampling One Error in...	48-kHz Sampling One Error in...	96-kHz Sampling One Error in...
10^{-12}	2 days	2 days	1 day
10^{-11}	5 hours	5 hours	2 hours
10^{-10}	30 minutes	27 minutes	14 minutes
10^{-9}	3 minutes	3 minutes	1 minute
10^{-8}	18 seconds	16 seconds	8 seconds
10^{-7}	2 seconds	2 seconds	1 second
10^{-6}	6 per second	6 per second	12 per second

Birth of the AES3 Standard

Our consideration of frequencies goes beyond the sampling rate. That is not what is traveling along the wire when we send digital audio. The sampled audio is assembled into "words."

The actual bandwidth is the sampling rate multiplied by 128. Table 4-4 shows our original sampling rates multiplied by 128. This then is the actual bandwidth that will appear on our cable. All our considerations and calculations will be based on these bandwidths.

If you remember one thing about this table, it is that every bandwidth is in megahertz. The limit of our hearing and therefore of analog audio is approximately 20 kHz. Now we are up to 25 MHz, more than a thousand times higher.

It cannot be so surprising then that designing a cable for digital audio will have a number of factors that weren't considered for analog audio. First is impedance, which is not important until the cable exceeds a quarter wavelength. Table 4-5 shows you the wavelength and quarter wavelength for the different digital audio bandwidths.

TABLE 4-4   Sampling Rates and Occupied Bandwidth

Sampling Rate	Actual Bandwidth
32 kHz	4.096 MHz
38 kHz	4.864 MHz
44.1 kHz	5.6448 MHz
48 kHz	6.144 MHz
88.2 kHz	11.2896 MHz
96 kHz	12.288 MHz
192 kHz	24.576 MHz

TABLE 4-5   Sampling Rates and Wavelength

Sample Rate (kHz)	Bandwidth (MHz)	Wavelength (ft)	Quarter Wavelength (ft)
32	4.096	240	60
38	4.864	202	51
44.1	5.6448	174	44
48	6.144	160	40
88.2	11.2896	87	22
96	12.288	80	20
192	24.576	40	10

The quarter-wavelength cable carrying 20 kHz, even with a poor plastic dielectric, was over a mile. So, for analog audio, cable impedance wasn't even considered. Any cable impedance was fine.

With digital audio, even the lowest sampling rate and lowest bandwidth produces a quarter wavelength of 60 ft. Can you have a cable that is 60 feet long? Of course you can. At the highest sampling rate and bandwidth, we're down to a mere 10 ft. The distances shown are wavelengths in "free space" and we haven't considered the dielectric that reduces the critical length even more.

In sum, impedance of the cable is important for digital audio and we must match the cable to the source and destination for maximum signal transfer. So what will the impedance be?

Balanced Line Twisted Pair AES Cables

Impedance was one of the problems that the AES/EBU committee had to solve. It was well known that 150 Ω was the best impedance for a twisted pair because it offers the lowest signal loss, or lowest attenuation (sharp readers will note that 150 Ω is double 75 Ω, the lowest loss for coax), and allows the signal to go the maximum distance. Impedance also determines the ratio of sizes and dimensions and, ultimately, the diameter of the cable.

If they chose 150 Ω as the standard and a manufacturer used a reasonably sized conductor, say 22 AWG, with a good dielectric like solid polyethylene, the resulting twisted-pair cable would have a diameter of approximately 0.310 in, almost a third of an inch. This is very big, too big certainly to fit into the XLR connector that the committee wanted to use and that would force cable manufacturers to use foamed dielectrics and much smaller conductors, possibly 26 AWG or even 28 AWG. There would be serious questions about ruggedness and reliability with conductors that small, especially when used on the road.

For these reasons the committee compromised on impedance to reduce size. By the time they were at 110 Ω, the cable was a reasonable size. This allowed a digital interconnect pair, with 24-AWG conductor foamed dielectric, foil shield, and drain wire, to have a diameter of around 0.180 in. That's reasonable compared to the 22-AWG analog interconnect at a diameter of 0.135 in. and 0.180 in. can easily fit into any XLR-style connector.

AES Impedance Tolerance

One fact hidden in the AES3 specification is impedance tolerance. It is not listed with the cable specifications but with the tolerance of source and destination devices. How far away can you get from 110 Ω and still have your system work? The answer is, quite far!

The tolerance the specification calls for is 110 Ω ± 20 percent. That means that a cable as low as 88 Ω or as high as 132 Ω would work. If we take the range of 88 to 132 Ω as gospel, there are many types of cables that fit in that range. Table 4-6 shows some of them.

TABLE 4-6   Potential AES Cables

Cable Type	Impedance (Ω)
RG-22B/U	95
RS-422	100
Twinax	100
Category 3	100
Category 5	100
Category 5e	100
Category 6	100
DeviceNet®	120
RS-485	120
SCSI cable	120
SCSI 2 cable	120
Twinax	124

You've probably never heard of many of these cables because they are used to control industrial systems or as premise/data computer cables. How many of these could you use to run digital audio? Surprisingly, all of them! Some of these industrial cables are huge or expensive. But would they work? Well, after all, these are almost all-data cables and digital audio is data!

Category cables are special cases. They meet all but one requirement: they do not have individually shielded pairs. Category cables are most often unshielded twisted pairs (UTPs). The AES3 specifically calls for shielded pairs. So can you use these unshielded cables? The answer is a tentative "yes" and we will explore the possibilities further on.

How Far?

Table 4-7 shows how far you can go on 110 Ω digital audio pairs. This information is based on the gage of the wire. These distance are the "safe" distances from the digital cliff.

TABLE 4-7   Digital Distance on Twisted Pairs by Gage

AWG	4 MHz	5 MHz	6 MHz	12 MHz	25 MHz
26	669 ft	602 ft	568 ft	434 ft	262 ft
24	793 ft	635 ft	599 ft	572 ft	312 ft
22	1094 ft	1048 ft	968 ft	786 ft	640 ft

Courtesy of Belden Electronics Division www.belden.com.

The AES specification requires a received voltage of at least 200 millivolts (mV) from a 5-volt source. It is easy to calculate, based on the resistance of a given wire, when that 200-mV minimum will be reached.

What difference does gage size make in the megahertz range, where the signal begins to travel on the skin of the conductor? The answer is, a lot! Even though skin effect is significant at these frequencies, it decreases as the wire gets smaller. Therefore, a larger wire produces less loss, no matter what the frequency, just because it has more skin.

Unbalanced Digital Audio

There are two unbalanced coax-based digital audio standards. The first is the professional version, called AES3-id. The second is the consumer standard, called the Sony/Phillips Digital Interface (S/PDIF).

AES3-id

This standard has been added only recently to the AES/EBU specifications. However, many engineers have been using coax to carry digital audio for more than a few years. In some ways, coax is superior to twisted pairs. Table 4-8 compares them.

TABLE 4-8   Digital Audio Cables

	Gage Size (AWG)	Impedance (Ω)	Impedance Tolerance	Capacitance (pF/ft)
AES Twisted Pair	26–22	110	$\pm20\%$ maximum, $\pm10\%$ typical	13
Coax	23–14	75	$\pm5\Omega$ maximum, $\pm3\Omega$ typical	20
Precision Coax	23–14	75	$\pm1.5\Omega$ maximum, $\pm0.75\Omega$ typical	16

There are three differences between twisted pairs and coax, some of which you already know. The first is balanced lines (twisted pairs) versus unbalanced lines (coax). Twisted pairs in a balanced line format can reject noise and interference. Surprisingly, this is not as big an advantage as it was for analog audio.

Analog audio dealt with very small signals, down to –60 dB and even lower. Any noise or interference that got by the common-mode rejection of the balanced line was virtually impossible to remove once it became part of the audio signal. Thus, anything that protected the analog signal was critical to good system performance.

Remember how I said at the start of this chapter that you need to stop "thinking analog"? This is one of those times. Digital signals are not analog. They are ones and zeros. If noise gets to a digital signal, it is easy to filter the noise out because it is not a one or a zero. Therefore, the noise rejection attributes of digital signals are amazingly low.

Whereas –90 dB of noise rejection was required for analog audio, digital requires only –30 dB. In fact, there is no standard in digital audio systems for noise or crosstalk. Many design engineers say that integrated circuits (chips) have become so good that the signal-to-noise ratio for digital systems can be as low as –3 dB, and many chips will work just fine.

Noise and Bit Errors

Incredibly, this improvement in technology means that the noise and interference can be almost the *same level* as the digital bits, and today's best chips can extract the bit stream from the noise with zero errors. So, although the tables that follow show a –30 dB noise requirement for digital systems, be aware that this number is very conservative and will get even more so as chips improve.

When noise affects a digital signal, it obscures the transitions, that is, the changes from zero to one or from one to zero. These transitions contain the clock, by which all digital audio is extracted. Changing the clock extraction is called jitter and can lead to misinterpretation of the state of a bit. These misinterpreted bits are then incorrect and called *bit errors.* The number of bit errors is directly related to the ability of the receiving device to reinterpret the bit stream and eventually turn it back into analog audio.

The improvement in chips and their ability to extract a perfect bitstream from a signal almost covered in noise is amazing and getting more amazing every day. This applies to all digital systems, not just digital audio or digital video but to computer systems and digital machine control systems.

Therefore, the distances shown in various charts and tables in this book should be taken as conservative "safe" numbers. The actual distance may be much greater. If you need to go farther than the distances outlined in this book, you will have to have some way of reading the bit errors. If there are no errors, then the cable works. Of course, you can always install a cable and then see if it works, but that's expensive.

What Coaxes to Use for Digital?

We have a problem right off the bat, as shown in Table 4-8. AES/EBU standards call for 110Ω twisted pairs. There are lots of coaxes out there, but none are 110 Ω and you wouldn't want one if there were.

Just like our twisted pairs, you want the lowest loss possible. In this case, we're in luck. The lowest loss coaxes are 75Ω coaxes, the very kind used for running baseband video signals. There are two kinds to choose from, standard and precision video.

AES devices that feed 75Ω coax usually use a transformer and a floating BNC, much like 10Base2 Ethernet. Floating the BNC helps with noise rejection because there is no common ground to other inputs and outputs. Even so, twisted pairs are superior in noise rejection.

How Far Can Digital Run on Coax?

Table 4-9 shows the distances you can run at various bandwidths. These distances are dramatically farther than with twisted pairs as shown in Table 4-7. However, these are safe distances, about half-way to the digital cliff. Specific precision video cables were used for Table 4-9.

The reasons for the amazing performance shown in Table 4-9 says a lot about the difference between twisted pairs and coax. First, the coaxes contain a lot more copper than two 24-AWG wires, rather common in a twisted pair. The losses on a coax simply mean that it will take longer to reach that 200-mV minimum AES signal strength. Second, the design of coax resists the movement of the components, resulting in an impedance tolerance that is three to five times better than that of a twisted pair.

What you lose going to coax is the balanced line and noise rejection of a twisted pair, but digital systems are very noise resistant, requiring only –30 dB of noise rejection or even less. Therefore, it will not surprise you to learn that there are a number of audio facilities throughout the United States wired up entirely with coax.

Digital Unbalanced versus Balanced

Just like unbalanced analog audio, coax for digital audio, also being unbalanced, lacks the inherent noise rejection of digital twisted pairs. The noise on the two conductors in a coax cannot be balanced because one conductor is at ground potential and the other is not. The only partial advantage to coax is a digital signal's natural rejection of noise, as shown by the –30-dB crosstalk requirement. If noise does appear on a digital coax, it will obscure the clock, which is embedded with the data. In other words, noise will add to the *jitter* of the signal, the inability to accurately extract the clock.

TABLE 4-9 Digital Audio Coax Distance

Precision Video	Diameter	Center conductor (gage)	4 MHz	5 MHz	6 MHz	12 MHz	25 MHz
Miniature coax	0.159 in	23	1527 ft	1368 ft	1253 ft	968 ft	699 ft
RG-59 coax	0.235 in	20	2188 ft	1997 ft	1831 ft	1398 ft	968 ft
RG-6 coax	0.275 in	18	2632 ft	2330 ft	2143 ft	1573 ft	1258 ft

In many high-frequency applications, noise is often ignored because it is so far removed by frequency ("out of band") compared with the signal. Most common noise sources are in the band used for digital audio. So, the simple rule would be, if you want noise rejection, choose a balanced line, a twisted pair. If you want distance, choose coax.

Coax Considerations

There are a few considerations when using coax to run digital audio that have nothing to do with performance. First is the fact that most AES-based equipment uses twisted-pair balanced line. Converting from balanced twisted-pair to unbalanced coax and from 110 to 75 Ω requires a balun. These are made by a number of manufacturers and are readily available, as shown in Figure 4-1. They cost between $50

Figure 4-1   Digital audio baluns.
(Courtesy of ETS, Inc. www.etslan.com.)

and $80, or even more, for each end. The price for two baluns, $100 or $160, will buy you a lot of cable. If you must convert a number of devices, these costs can really add up.

WHAT DO I DO NOW?

I remember talking on the phone with a customer. He had ordered the latest, hottest mixing console from England for his new all-digital studio. When it arrived, he was shocked to see the entire back was all BNCs, and used coax cable entirely. Unfortunately, the studio he had already built used only 110Ω twisted pairs. "What do I do now?" he pleaded. He could have called the console manufacturer, but the alternate console was 50 percent bigger and he had designed to have this smaller console "drop" in his studio design. He could have bought 100 baluns ($5,000 at least) and it would have been a mess because the baluns were bigger than BNCs and wouldn't all fit on the console. He was in a no-win situation.

Coaxes connect with BNC connectors. These connectors have a very small "footprint" especially compared to XLR and other connection schemes for digital twisted pairs. If compactness, size, and weight are considerations, coax-based mixing consoles, for instance, may be too good to pass up.

Just be sure you make a conscious choice between coax and twisted pairs; otherwise, you may be in a pickle (see previous box). The dimensions, compactness, and weight of BNC-based consoles may lend themselves to audio or video production trucks and ENG/EFP vehicles. This choice allows you to use the same cable for audio and video, a very cost-effective solution. The only concession might be picking a different color cable for audio than for video.

One important consideration with baluns such as those shown in Figure 4-1 is just how much depth they can add to a unit. It is wise to consider not only the machine itself but also the baluns that may be attached and an appropriate bend radius for the cable from there on. The balun connector and cable together may add more than 5 in. to the effective depth. Those inches may be especially critical in installations where rack size and depth are critical to the inch, such as in trucks. We'll look at trucks and different forms of digital installations in Chapter 9.

1-Volt AES Baluns

There are balun types other than those shown in Figure 4-1. They contain a resistive network to drop the voltage from 5 volts to 1 volt. This product is targeted to television stations currently converting to digital video and AES digital.

The AES signal at 1 volt is the same voltage as the video signal. Instead of discarding those old patch panels for analog video, you can use them for patching AES in the coax format.

Further, many old analog video distribution amplifiers can be used with digital bit streams. These amplifiers require a 1-volt input. Whether your D/As will work up to 25 MHz (192-kHz sampling) is beyond the scope of this little book. You can certainly run some AES through the amplifier and look at the bit errors or contact the manufacturer and ask whether your unit will perform with this new application.

The other advantage is that you can use one cable for both digital audio and digital video. Perhaps you might select a different color, so you can recognize which cables are video and which are audio. But the possibility of using one cable, one strip tool, one crimp tool, and one connector is hard to resist.

S/PDIF

S/PDIF is the consumer version of digital audio. It is often pronounced *spid-if* or *es-pee-dif*. It is and has always been based on coax. It also is based on the RCA, or phono, connector. This connector is certainly much less than 75 Ω. The quarter wavelength of a 44.1 kHz sampled CD is 44 ft. Obviously a connector that is an inch or shorter would have no effect. It is unlikely that anyone would want to run digital audio in a home installation for 44 ft, but someone might.

Most cables are 1 or 2 m (3 to 6 ft), too small to make a difference. However, if you want to run that AES signal to another room, maybe at the end of your home, 44 ft suddenly seems to be a reasonable length.

What is the impedance of a standard hi-fi interconnect? Nobody knows. The source and destination impedances in these analog devices are 75 Ω, so why not use video cable for digital audio? Why not indeed? But this is where a lot of do-it-yourself home installers go wrong.

First, they assume that "video cable" is the same as *cable television cable*, also called CATV/broadband cable. CATV cable is not video cable but RF cable! It is designed to carry multiple television channels. It isn't designed to start working correctly until 50 MHz. (Channel 2 is 54 MHz.) The highest possible AES bandwidth is less than 25 MHz, so you can see this is a poor choice. What they really want to use is *baseband video* or, even better, *precision video*. In truth, any cable intended to carry baseband video (4.2 MHz) will work just fine for digital audio, and those "precision" cables may be overkill. But nobody was ever fired for buying the best there is. We'll discuss CATV/broadband cable in Chapter 7.

Are there RCA connectors for the cable you are considering? If not, you will end up putting on a BNC connector because video cables standardize on BNCs. Then you will buy an adapter to convert a BNC to an RCA. This is a kluge,

at best, and adds to the unreliability of the system. It's better to select a cable that fits into a standard RCA or one that
has RCAs made specifically to fit it.

One problem that is rarely mentioned is the distance limitation in S/PDIF. Its source voltage is only 0.5 volts, a tenth
of what the professional AES3-id delivers. This means that
S/PDIF cannot go as far as AES3-id, but that's fine because
we're talking thousands of feet (Table 4-9). Even at one-
tenth the distance, we're still at totally acceptable distances
for even the largest home.

Can I Run Digital on Analog Pairs?

What about the billions of feet of analog cable already
installed? Why can't it be used to run digital systems? Well,
it can, with some severe restrictions. Table 4-10 shows the
requirements of a digital system and the parameters of analog cable.

TABLE 4-10   Digital versus Analog Cable

	AWG	Impedance	Capacitance
AES requirements	26–22	110 Ω, 20%	13 pF/ft
Analog cable	26–22	30Ω–70 Ω	30–50 pF/ft

Table 4-10 shows that analog pairs are less than ideal in
two areas, impedance and capacitance. The low capacitance
of the digital cable is a result of the choice of impedance.
Because impedance determines the relation between the
parts of the twisted pair, the ratio between conductor size
and spacing, the capacitance is a natural reaction to those
dimensions. For instance, a 150-Ω AES cable would have a

capacitance of 8.5 pF/ft. Too bad that impedance makes a huge, and unusable, cable!

Figures 4-2 to 4-4 show what would happen with analog pairs. In all three, we introduce a 48-kHz (6-MHz bandwidth) AES digital signal, shown as the solid square lines in the graphs, to two different twisted pairs.

The first cable, the solid resultant line, is analog cable, 22 AWG, 34 pF/ft with an impedance of 38 Ω. On the second cable, the dotted resultant line is also 22 AWG. We must maintain the same gage size; otherwise, this would not be a fair comparison. This second cable is 110 Ω, 13 pF/ft intended for digital audio.

Figures 4-2 to 4-4 show the resultant signals at three different distances. Figure 4-2 shows two waveforms resulting from the two cables at 50 ft. Even at this short distance,

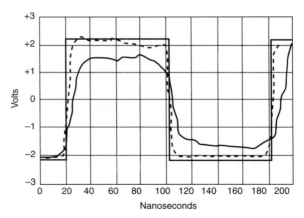

Figure 4-2 Digital signal on analog cable at 50 ft.

there are noticeable differences. Both signals would most likely be recoverable.

Figure 4-3 shows the same two cables at 100 ft. Now we have some serious differences.

The signal on the analog cable has decreased from 2 volts to 1 volt, a 50 percent (−3 dB) drop. That is due to the impedance mismatch. The digital device wants 110 Ω but gets 38 Ω. This mismatch causes return loss, where the signal reflects back to the source instead of getting to the destination. In this case, at only 100 ft, half our signal is reflected.

The other problem, capacitance, shows up at the edges of the waveform. As it rises from zero to one, and remains there before the one returns to zero, this square wave is really a very complex pattern. The rise from zero to one is supposed to be instantaneous and is therefore an "infinitely" high frequency. The horizontal excursion until the next transition is

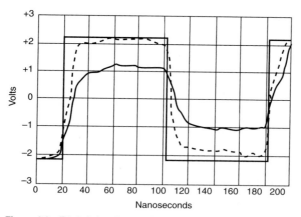

Figure 4-3 Digital signal on analog cable at 100 ft.

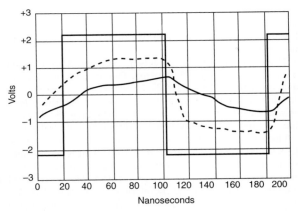

Figure 4-4 Digital signal on analog cable at 300 ft.

really a very short DC pulse. And DC, as you probably have guessed, is no frequency, or an "infinitely" low frequency.

If you know anything about square waves, you know that they are made up of an infinite number of harmonics around the clock frequency. *Harmonics* are multiples of the original frequency. Therefore, the better the cable, with lower capacitance, the more harmonics get through and the more square the square wave will be.

That is the reason analog cable at 34 pF/ft has difficulty in handling the harmonics and high frequencies of the square wave. In fact, you can see the capacitance of that cable "charging up" on the voltage of the signal and then "discharging" when the signal goes from one to zero.

This effect of "rounding the edges" of the signal is deadly in digital systems. Because the clock is embedded in the digital signal, high-capacitance cable "blurs" exactly where the transition points are, making it difficult to accurately recover the clock signal. Those exact points, where zero becomes one and one becomes zero, carry all the digital information. If you could invent a system that passed on only those points, you would have a perfect digital system. That's your billion-dollar idea for this week!

Effects that blur the transitions and thus affect the ability to extract the clock are called *jitter*. Don't let anyone tell you that cable cannot cause jitter. Jitter can be caused in part by a poor cable choice for running digital. The cable also has to be long enough to show the effects.

Figure 4-4 shows the two cables at 300 ft. The analog cable has become completely useless. No information can be extracted from that signal. The digital cable is a lot better, but it's not perfect. After all, this cable doesn't have zero capacitance, it has 13 pF/ft, about a third of the analog cable. The impedance tolerance of this cable design is about 110 Ω ±10 percent, so there is some small impedance mismatch. It is not exactly 110 Ω.

You can now see that coax works much better in the "distance wars" of digital audio. The impedance tolerance of a high-quality precision video cable is ±1.5 Ω maximum and typically ±0.75 Ω. That's eight times better than the twisted pair shown.

Isn't it too bad that we can't make a twisted pair with impedance tolerance like a coax cable? Well, stick around, because we can. We'll be getting to that in our discussion of category cables on the next section.

Can I Run Analog on Digital Pairs?

You may recall from Chapter 3 that the key parameter for judging the performance of an analog cable is capacitance. The capacitance of these digital cables is 13 pF/ft. That's two to four times better than the standard analog cables.

Therefore, digital cables make the best analog cables ever made.

If you have customers who want no-holds-barred performance for analog, put in digital cable. Not only will they have unmatched performance, but they'll be wired and ready to go when converting to digital. They may even use the exact same connectors. This is called *future-proofing*.

Future-proofing

When you design a future-proofed analog system, design it as if it were a digital system. Be aware of and remember the length limits for digital. They don't apply to analog audio, but they will apply when your customers convert part or entirely to digital. They will be your most satisfied customers because you did two installations for the price of one.

Those customers who balk at using digital audio cable because it is two to three times more expensive than analog just don't realize what they're getting for that money. For one thing, the cost of the wire and cable in most installations is 5 to 10 percent of the total cost of the installation. Even so, the largest single cost item is often the labor to put in that same cable. If that customer wants you to come back in a few years, rip out the old analog wiring, and put in new digital cable, well, I guess that's job security for you! If your customer balks because it says "digital" on the cable and you're running analog, I will personally send you a bottle of rubbing alcohol so that you can rub the word *digital* off the cable so that your customer won't be so confused.

Just remember that in future-proofing we're talking about putting in AES/EBU twisted pairs and running balanced line analog audio down them. The same applies to patch panels. If you install low-capacitance digital audio versions, they can only improve your analog performance now. They will pay for themselves when you change to digital because you will not have to replace them or even move them. Future-proofing like this only applies to twisted pairs and balanced lines in the professional audio world.

In home theater installations that use unbalanced interconnects, the same idea applies. By using video coax, especially precision video coax, you can future-proof that customer. Of course, the length limits in digital are so long and most interconnect cables for home applications are so short that this is an easy step to take.

Return Loss and Digital Audio

Return loss is caused by impedance mismatching. It can apply to various parts of a circuit, such as connectors or patch panels, or to the cable itself. Return loss shows the variations in impedance or the loss due to the use of cable with an impedance far from the desired value.

With return loss, a percentage of the signal will reflect, or return, to the source. The severity of the problem is directly related to frequency. At low audio frequencies, where the impedance of the cable doesn't matter, neither does the impedance mismatch of the actual cable impedance to the source or destination impedance. It does matter when the interconnect length is a quarter of a wavelength or longer.

What is bad return loss?

Table 4-11 shows a range of return loss values and whether those values are good or bad. This is a rough scale to give you a general idea of what is good or bad return loss. The lower the number (the larger the negative number), the better the performance (the less the return loss).

TABLE 4-11   Return Loss

Value	Performance
0 dB	Inoperative
−10 dB	Very bad
−15 dB	Poor
−20 dB	Good
−25 dB	Very good
−30 dB	Excellent

Be aware that significant return loss, say −15 dB, does not mean that your circuit won't work. It just means that you are wasting your signal strength for no reason. If it is a short interconnect, you probably won't even notice the problem. In long connections, however, it might be the reason you get excessive bit errors, or even nothing at the end of the cable, because your signal is now below the 200-mV threshold of most digital devices.

At digital audio frequencies, we might install cables a quarter of a wavelength or longer. Does that mean using cable that is not exactly 110 Ω causes return loss? Yes and no. It depends on how far apart the cable impedance is from the desired value. Figure 4-5 shows a simple formula to determine return loss from two values of impedance.

RL is the return loss in decibels. Difference is the smaller value subtracted from the larger one. The sum is the two values added together. Table 4-12 shows return loss values for various common analog and digital cable impedances versus the desired value, 110 Ω.

$$RL = 20 \log \frac{DIFFERENCE}{SUM}$$

Figure 4-5   Formula: return loss.

TABLE 4-12   Other Cables versus 110-Ω Digital Cable

Cable Impedance (Ω)	Return Loss (dB)
30	−5
40	−7
50	−9
70	−13
75	−15
88	−19
95	−23
100	−26
120	−27
124	−25
132	−19
150	−16

Table 4-12 explains a lot of things. For instance, in Figures 4-2 to 4-4, which show what happens when digital signals run on analog cable, there was increased capacitance and return loss. I used the term "impedance mismatch" then, but that was really return loss. That's what caused the resultant signal in the analog pair to drop in level compared to a digital cable. In Figure 4-3, at 100 ft, half the signal was reflected back to the source. Because those analog cables are generally 30 to 40 Ω, we're talking serious return loss, as shown in Table 4-12.

You'll also notice that the AES range of 88 to 132 Ω for paired cables produces a return loss of slightly over −19 dB, or "good." You might think, with a 192-kHz sampling rate taking us to bandwidths of 24.576 MHz, that this range should be tightened in the spec.

The truth is, cable manufacturers are having a hard time just staying within the ±20 percent requirement. The best twisted-pair cables made specifically for digital audio are around ±10 percent. Can we do better than ±10 percent? Not really, and there are a number of reasons why it's difficult.

First, these cables use stranded conductors. The exact shape and position of each conductor change when the cable is moved around. Also, the two conductors, while twisted together, also move and separate when the cable is handled. The exact dimensions and distance between are the key to the precise impedance. Therefore, these cables have inherent impedance variability.

Is it possible to build cables that reduce these effects? Certainly, and that will take us to some of those category cables in a few pages.

Table 4-12 shows that, as you get closer to the correct value, the return loss gets lower and lower. Once you pass the correct value, the return loss beings to increase again.

Table 4-12 also explains the occasional comment I get from installers who say, "I just hooked the 75-Ω coax up to the 110Ω input and it worked just fine..." or vice versa. Well, it's not just fine, but it's not deadly either. You have a significant return loss, almost −15 dB, and that will eat up a bit of signal level, but it probably will work just fine, especially if your cable is short. Of course, if you used what was intended, you would have almost no return loss at all.

I have to say "almost no return loss" because there always are variations in impedance. You may recall from the beginning of this section that the AES specifications for twisted-pair cable call for an impedance of 110 Ω ± 20 percent. That means the cable (or the input or output of a digital device) could be anywhere from 88 to 132Ω and still meet this specification. Even a cable with "110 Ω" printed on the jacket might have some return loss. Ultimately, you would include the inherent variations in impedance caused by stranded conductors and variable conductor spacing.

So it is even more obvious why coax wins in terms of return loss, because the typical impedance variations on coax are dramatically less than impedance variations on twisted pairs. The world's worst 75Ω video coax might be

±10 Ω, but standard quality video coax is generally ±5 Ω maximum and ±3 Ω is typical. Precision video coax is ±1.5 Ω maximum and ±0.75 Ω typical. You can see that the lower the possible variation in impedance, the lower the return loss. With lower return loss, you get greater signal strength and longer cable runs. Table 4-13 proves it.

TABLE 4-13   Return Loss of 75-Ω Coaxial Cable

Maximum Variation (Ω)	Return Loss (dB)
75 ± 10 (65, 85)	−23
75 ± 5 (70, 80)	−24
75 ± 3 (72, 78)	−28
75 ± 1 (74, 76)	−38

The formula in Figure 4-5 calculated the return loss difference by using the difference between impedance values, so it doesn't matter whether the cable impedance is too high or too low. In other words, the return loss in the first example (75 ± 10 Ω) doesn't change, whether the cable is 10 Ω low (65 Ω) or 10 Ω high (85 Ω); the difference, compared to the desired value, is still 10 Ω off.

Note that even the "bad" coax does very well. At a return loss of −23 dB, it would rate between good and very good in Table 4-11. The real problem with cheap coax is that the manufacturer never measures the impedance variations (much less the return loss) on its cable.

This should motivate you to find out what the return loss values are for the cable you are purchasing. If the manufacturer cannot furnish you with these values, then the cable may not be the highest-quality product. This will become especially significant when we discuss digital video cable.

Splitting Digital Audio

I would bet that you already know what the problem is with splitting digital signals. It is return loss. In analog audio, splitting signals resulted in a level loss (Table 3-14) but nothing more.

With digital audio, every time we lay one pair on the same terminals as another pair, or buy a "T" splitter for coax, we are putting two cables in parallel. The problem is the resulting impedance. Table 4-14 shows the return loss when adding multiple cables in digital audio.

TABLE 4-14   Splitting Digital Signals and Return Loss

	110Ω Cable		75Ω Cable	
Number of Cables	Resulting Impedance (Ω)	Return Loss (dB)	Resulting Impedance (Ω)	Return Loss (dB)
1	110	—	75	—
2	55	−9.54	37.5	−9.54
3	36.7	−6.02	25	−6.02
4	27.5	−4.44	18.75	−4.44

After only one split, the twisted pair and coax are below "very bad" as rated by Table 4-11. So how do you split cables? You use an impedance-matched splitter. For digital audio, these are available in the coax format, with BNC connectors (Figure 4-6), and the balanced line 110Ω format, with XLR connectors (Figure 4-7).

Connectors

Don't you need impedance-specific connectors for digital audio? You don't and the reason is quite simple. Look at

Figure 4-6 Coaxial digital splitter.
(Courtesy ETS, Inc. www.etslan.com.)

Table 4-5 and you will note that the quarter wavelength of highest bandwidth for digital audio is 10 ft. Then what effect can a connector have? Pretty much, none!

This is the reason, in the AES twisted-pair standard, they call for the ubiquitous XLR-style connector. These three-pin connectors have been the standard since the 1960s, are easily obtained, and come with a wide range of options, a wide range of quality, and an equally wide range in price.

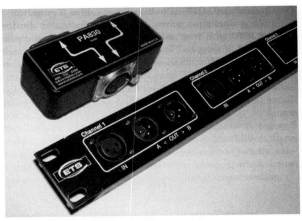

Figure 4-7 Twisted-pair digital splitter.
(Courtesy ETS, Inc. www.etslan.com.)

Patch Panels

Patch panels, when used for digital audio applications, have a number of added considerations. Both common sizes, the TRS and the bantam, are not impedance specific. These connectors are also in the 30- to 40-Ω range. However, because the critical distance is at least 10 ft and a quarter wavelength of 24.576 MHz (Table 1-13), a few inches of connectors will make little difference.

That being said, there is probably measurable return loss simply because of the abrupt transition from 110 to 30 Ω. Abrupt transitions are more likely to affect high frequencies than gradual changes. If you intend to use the same historic types of patch panels, you won't have much choice.

Some patch panel manufacturers have versions that are lower capacitance. As shown in Figures 4-2 to 4-4, high capacitance is a significant drawback when using standard analog cable to run digital. Connector manufacturers can control capacitance so that a low capacitance version would be preferable for digital audio applications.

Most audio patch panels are set up with *normals*. Normals allow signals to flow in and out of jacks, sending signals where they would "normally" go. You insert a patch cord only when you wish to change the source or destination. In most double-row patch panels, the top jack is the source and the bottom jack is the destination.

Inserting a patch cord in a source jack disconnects that source and allows you to send its signal to a different destination. But the destination jack is often half-normaled, meaning that insertion of a patch cord will not disconnect the destination, but rather mult, or add on, the patch cord as a second place to send that signal.

Half-normaling, where signals can be split two ways, are, of course, unacceptable for digital, as can be seen by the terrible return loss caused when digital signals are split (Table 4-14).

There are companies that make rack panels with multiple digital splitters. These devices are passive impedance-specific splitters, with one input and two outputs. They can be daisy-chained to provide more outputs, but every split drops the signal 3 dB (not including the insertion loss of the device), so you can only do that so many times before you have a very weak digital signal.

Digital audio patch cords

Because twisted-pair digital cable makes the best balanced analog cable, it makes sense to be sure that the patch cords are also 110Ω digital audio cable. They are probably too short to make a difference, but seeing that they are always part of a chain, and that chain can easily be long enough to make a difference, changing the patch cords is advisable.

S/PDIF, AES3-id, and patch panels

If your installation is in the coax mode, you have some very interesting options available in terms of patch panels. For one thing, impedance-specific 75Ω patch panels are the norm. There is no half-normaling in analog video, so that's no problem. The patch cords also are 75 Ω. So you will get little or no added return loss there.

If your installation is part of a video facility, you have one very clever option. If your video is going digital or high definition, you will have to change out your old patch panels for ones that perform at those high frequencies. But don't toss those old video panels. Use them for digital audio. They are perfectly acceptable for even the highest digital audio bandwidth.

Digital Audio Distribution

If you used distribution amplifiers to send the old analog video around, you may be able to use them to function as digital audio distribution amplifiers. Just check in the service manual or with the manufacturer to see if those distribution amps have enough bandwidth and a good square-wave response to handle digital signals. You could be off and running!

You also can use the same cable you've chosen for serial digital or high definition to carry digital audio. This means one kind of cable, one connector, one strip tool, one crimp tool, and no confusion. You might choose a different color of cable to indicate digital. In Europe, purple means digital, so there's a suggestion. We'll talk more about this one-cable solution in Chapter 6.

Category 5 and Digital Audio

If you've been reading this book in chapter order, you've seen all the hints that this was coming. If you can use Category cables to carry analog (pages 127 to 135), then why not digital? The "100 Ω" line in Table 4-12 shows a return loss of −26.44 dB, which is rated as "very good" in Table 4-11.

The capacitance of this cable is very low, 15 pF/ft, and this cable is designed specifically for digital signals. Sure, it's computer cable, but those are also digital signals. Table 4-15 compares the requirements of twisted-pair balanced line digital audio to the specifications of Category 5, 5e, and 6 cables.

From Table 4-15, it is clear that most data cables are solid conductor, whereas most audio cables, including digital audio, are stranded.

TABLE 4-15   Category Cables and Digital Audio

	AES Spec	Category 5	Category 5e	Category 6
Impedance	110 Ω ± 20% (88–132 Ω)	100 Ω ±15 Ω (85–15 Ω)	100 Ω ±15 Ω (85–15 Ω)	100 Ω ±15 Ω (85–115 Ω)
Capacitance	<20 pF/ft	15 pF/ft	15 pF/ft	15 pF/ft
NEXT				
6 MHz	30 dB(?)	51 dB	54 dB	63 dB
25 MHz	30 dB(?)	50.5 dB	53.5 dB	62.5 dB
Gage	22/24/26	24 AWG	24 AWG	24 AWG
Format	Stranded	Solid	Solid	Solid

Solid versus Stranded

Table 4-16 shows the differences between solid and stranded cable. Many of these differences apply to analog audio and digital applications. Others are more critical at higher frequencies or data rates, such as computer cables. By the time you're at those frequencies (100 MHz), there's a separate standard for stranded cables, as it is dramatically worse than solid conductors.

TABLE 4-16   Stranded Versus Solid

Parameter	Stranded	Solid
Flexibility	Superior	Good
Flex-life (flexes to failure)	Superior	Poor
Low-frequency performance	Equal	Equal
High-frequency performance	Average	Good
Very high-frequency performance	Poor	Excellent

The main question is ruggedness. If you're going to use this cable on the road, stranded will last longer than solid. If you're going to install it, flexibility or flex-life doesn't matter, In fact, as pointed out in the previous section on snake cable (page 116), flexible cables are sometimes harder, not easier, to install.

Shielded or UTP?

Table 4-15 showed that all the performance requirements are easily met. There's really only one that isn't and that is shielding. Most premise/data computer cables are UTPs. The AES specifically calls for "shielded" cable. So what can you do?

Who owns the spec?

If you are building a facility and your boss or customer has said, "This facility will follow the AES3 guidelines and specifications," then you are going to use shielded digital audio cable. However, if your boss or customer said, "Just make it work," then you're pretty much in control. If it's your own facility, then you are the keeper of the spec and you can make it work.

Category 5 and Return Loss

There's one other thing you might consider. Category cable, like any other, is not absolutely precise. It is not exactly 100 Ω, but it has specified impedance, which is 100 ± 15 Ω. That means it could be anywhere from 85 to 115 Ω.

If you compare that range to the digital audio spec, 88 to 132 Ω, you have some possible impedance mismatch. Table 4-17 shows all the possible combinations. Be aware that these numbers can represent cable or the inputs/outputs of a specific device. To be sure, it is more likely that a piece of cable will vary more in impedance than the input or output of a device. Therefore, the numbers in Table 4-17, where the 110 Ω is stable but the Category cable is not, are probably the commonly expected combination.

You can see that the return loss numbers are all over the place, from "poor" to beyond "excellent." It's obvious that when the excursions are in the same direction, return loss can be very good. When the excursions are in opposite directions, you can get some very poor performance. What can you do? Call the cable manufacturer and ask them to make cable to a tighter tolerance? Well, they already have. The magic words are "bonded pairs."

TABLE 4-17   Category Cables, Digital, and Return Loss

Category cable impedance	AES cable impedance	Return Loss (dB)
85	88	−35
85	110	−18
85	132	−13
100	88	−24
100	110	−26
100	132	−17
115	88	−18
115	110	−33
115	132	−23

There are families of premise/data cable that use bonded pairs. In these cables, the conductors for each pair are extruded, twisted, and joined together in one step. There is no glue holding the conductors together. Because the impedance of a twisted pair is determined by the spacing between conductors (Figure 1-29), if you keep the conductors from moving around, you will reduce variations in impedance. These cables are specified as ±10 Ω maximum and typically ±7 Ω. As you can see, these bonded-pair data cables are approaching coax performance with all the noise rejection of a twisted pair.

Bonded-pair cables are excellent choices for digital audio installations. A bonded-pair Category 6, for instance, can provide over a thousand times more crosstalk protections than digital signals require (Table 4-15). So who needs a shield?

MADI

Formulated in 1991, MADI stands for *multichannel audio digital interface*. It is a standard of AES (AES-10-1991) and essentially the multichannel version of AES/EBU. It supports up to 56 channels of audio, with a separate sync signal. It uses a standard single coax cable with BNC connectors to carry all 56 channels. Its intention is to be the connector between multichannel consoles and recorders.

MIDI

Formulated in 1982, MIDI stands for *musical instrument digital interface*. MIDI employs a relatively slow data rate (31.25 kbaud) to interconnect computerized musical equipment such as synthesizers, audio processors (equalizers, reverbs, delays, compressors, limiters, and so forth), foot-pedal controllers, and samplers.

The signals running down MIDI cables are not audio. They are control signals that specify actions, such as the playing of specific notes on a synthesizer.

The cables usually consist of two twisted pairs. Data is carried as an asynchronous serial transmission on one pair. Because the data rates are in the audio range of bandwidth, there is no specified impedance and any twisted-pair cable will work perfectly well. The same limitations of audio cable apply.

The second twisted-pair could just as easily be two straight wires because it is intended to deliver power to external devices. Because this pair has a smaller gage, it is not intended to actually power up whole boxes but provide power to devices such as foot pedals. This would be enough power for a chip or two and sending a signal to control another device. "Control-power only," to coin a term.

If only data is traveling between devices, a single-pair shielded cable would work as well. The connector specified is a five-pin DIN connector, similar to the mouse and keyboard connectors used for computers although, with the low bandwidth, any connector (or even hard-wired connections) would work just fine.

Networked Audio

Finally, there is the emerging technology of networked audio. This approach makes abundant sense. Computer networks are made to carry data. They don't care what the data is. It can be a spreadsheet, email, or audio. The network doesn't care. Therefore, most of what you would need to know would be in a book about network architecture. (See Reference section at the end of this book.)

This is not digital audio as we have known it. It is audio transformed into a computer network protocol and extracted at the other end. It conforms to the computer network standard, not to AES or any other digital audio specification. Most of these systems use existing network architecture and systems such as Ethernet®. I will discuss one such system, Cobranet.

Cobranet

Cobranet is a system sponsored by Crest Audio, manufacturer of power amplifiers and similar equipment. The intention is to carry up to 128 channels of audio on one four-pair Category 5 cable.

This system is cosponsored by Biamp, Clear-Com, Creative Audio, Crown, Digigram, EAW, LCS, Peavey, QSC, Rane, Whirlwind, and a growing list of other manufacturers. If you know these manufacturers, you know their key market is sound reproduction, especially large-venue multichannel systems.

One main problem with Cobranet is simply running 128 channels down one Category 5 cable. What if the cable is cut? What if a connector fails? Instead of one microphone going dead, the entire concert or convention goes dead.

If the Category 5 cable is installed and protected, it is probably not a issue. But for a traveling show, protecting the cable becomes a cause of concern. If this is a grunge or heavy metal band on the road, the cables might run through the "mosh pit" on the way to the FOH mixer. That's a serious problem.

The solution is "ruggedized" Category 5e or 6 cables. These sometimes use double jacketing, or even interlocking steel armor, to protect the cable inside. The steel-armored version has a plastic jacket on the outside. That's as close to "mosh-pit proof" as one's going to get in a category cable.

Chapter 5

Analog Video

A Very Short History of Television

The National Television System Committee (NTSC), at its meeting on March 8, 1941, adopted the standard of a 525-line picture. They also accepted the specifications previously suggested in 1938, by the Radio Manufacturers Association (RMA) Television Standards Committee of specifications for brightness, horizontal polarization, synchronizing signals, and vestigial sideband transmission. This increased video bandwidth to its current 4.2 MHz. Previous work in 1936 had set the 4:3 aspect ratio and the 2:1 interlacing ratio that we presently use with analog video.

Very soon after, the war effort overshadowed any work being done in television. Live broadcasts, albeit experimental, had begun in the United States and Great Britain. We might have begun commercial broadcasting as early as 1942. But waiting might not have been such a bad thing. At the conclusion of the war, there were thousands of ex-military men trained in RF theory, transmission line design and installation, and antenna theory and design. This gave the television effort a major boost in 1946 and 1947, when it was restarted.

The military also produced cables made with the new "Top Secret" plastics, such as polyethylene (see page 14 for more on polyethylene). Future cables would also use the other "Top Secret" plastic, Teflon. Great strides also had been made in non-critical plastics such as PVC and polypropylene. The war effort, as war always does, helped the required technology to mature. So it is quite possible that the work required to put television on the air would have taken exactly as much time as it did take.

Which Cables for Video?

When commercial television restarted, the military had a long line of "standard mil-spec" cables to choose from. Table 5-1 shows a list of many common types of cable available at the end of the war. Choosing one rather than another was a simple matter of determining the lowest loss (Figure 2-8). The chosen cables were, of course, 75 Ω. You'll note a number of 75 Ω versions to choose from. Many technologically related descendent cables are available. Some of these still extant cables are older technology and are slowly disappearing.

The RG, or *radio guide*, number was a military system started in the 1930s. It was a book with specifications. So RG-8 and RG-9 were just one section away from each other. There was no other relationship.

So what happened to RG-1 or RG-65, or any other missing number? They undoubtedly proved to not be very popular. They were possibly special constructions intended for specific applications. Of course, there were "winners" in Table 5-1. Table 5-2 shows the winner's circle for video transmission.

TABLE 5-1   RG Coaxes

RG Number	Center Conductor		pF/ft.	Impedance (Ω)	Outer Diameter (in)
RG-5/U	16 AWG	Silver-clad copper	Solid	28.5	50	0.238
RG-5B/U[1]	16 AWG	Silver-clad copper	Solid	28.5	50	0.328
RG-6A/U	21 AWG	Copper-clad steel	Solid	20	75	0.332
RG-8/U[1]	13 AWG	Bare copper	Stranded	29.5	52	0.405
RG-8A/U[1]	13 AWG	Bare copper	Stranded	29.5	52	0.405
RG-9/U[1]	13AWG	Silver-clad copper	Stranded	30	51	0.420
RG-9A/U	13 AWG	Silver-Clad Copper	Stranded	30	51	0.420
RG-9B/U[1]	13 AWG	Silver-clad copper	Stranded	30	51	0.420
RG-11/U[1]	18 AWG	Tinned copper	Stranded	20.5	75	0.405
RG-11A/U[1]	18AWG	Tinned copper	Stranded	20.5	75	0.405
RG-13/U	18AWG	Tinned copper	Stranded	20.5	74	0.420
RG-14A/U	10 AWG	Bare copper	Solid	29.5	52	0.545
RG-17/U	0.188 in	Bare copper	Solid	29.5	52	0.870
RG-17A/U	0.188 in	Bare copper	Solid	29.5	52	0.870
RG-19A/U	0.250 in	Bare copper	Solid	25.9	52	1.120
RG-21A/U	16 AWG	Nichrome	Solid	29	53	0.332
RG-22/U	2–18 AWG	Bare copper	Stranded	16	95	0.405
RG-22B/U	2–18 AWG	Bare copper	Stranded	16	95	0.420
RG-34B/U	14 AWG	Bare copper	Stranded	21.5	75	0.630
RG-54A/U	18 AWG	Bare copper	Stranded	26.5	58	0.250
RG-55/U	20 AWG	Bare copper	Solid	28.5	53.5	0.206

continued on next page

TABLE 5-1 RG Coaxes (continued)

RG Number		Center Conductor		pF/ft.	Impedance (Ω)	Outer Diameter (in)
RG-55B/U	20 AWG	Silver-clad copper	Stranded	28.5	53.5	0.206
RG-58/U[1]	20 AWG	Bare copper	Solid	28.5	53.5	0.195
RG-58A/U	21 AWG	Tinned Copper	Stranded	28.5	50	0.195
RG-58C/U[1]	21 AWG	Tinned copper	Stranded	28.5	50	0.195
RG-59/U[1]	22 AWG	Copper-clad steel	Solid	21	73	0.242
RG-59B/U[1]	23 AWG	Copper-clad steel	Solid	21	75	0.242
RG-62/U[1]	22 AWG	Copper-clad steel	Solid	13.5	93	0.242
RG-62A/U[1]	22 AWG	Copper-clad steel	Solid	13.5	93	0.242
RG-62B/U[1]	22 AWG	Copper-clad steel	Stranded	13.5	93	0.242
RG-63/U[1]	22 AWG	Copper-clad steel	Solid	10	125	0.405
RG-63B/U[1]	22 AWG	Copper-clad steel	Solid	10	125	0.405
RG-71/U[1]	22 AWG	Copper-clad steel	Solid	14.5	93	0.250
RG-71B/U[1]	22 AWG	Copper-clad steel	Solid	14.5	93	0.250

[1]These cables are still available.

TABLE 5-2   Common RG Coaxes

RG Number	Common application
RG-6	Most common broadband/CATV, precision HD versions
RG-11	Low-loss, long-run CATV/broadband, precision HD versions
RG-59	Most popular analog video cable, precision HD versions

The RG system continued through the RG-400 series. It was replaced by the current "mil-spec" system. Part of the problem was that a large percentage of the RG numbers had been abandoned and those "dead" designs made the system unwieldy.

What about the As, Bs, and Cs, not to mention the Us? The "U" stands for "utility" grade, the standard grade of that cable. The A, B, and C designations indicated changes in construction. It might be a change from a solid to a stranded center, or to a special noncontaminating jacket compound. These changes often can result in a change of dimensions and, hence, a change in connector. As you can imagine, this creates chaos with a half-dozen connectors, all with slightly different dimensions, for cables that are essentially the same.

That leads us to the "RG Type" designation. Most catalogs list an "RG-59 Type" or some other RG number with the word *type*: "Kind of like the original RG-59 specification, but not really." It's like saying the cable is "in the family" of this original cable. Certainly, most modern RG-59s or any other modern coax bears only a superficial resemblance to the original cable of 50, 60, or 70 years ago.

For instance, all the current RG-6 cables have a diameter of 0.275 in. I doubt you could find any cable with the original .332 in. diameter as shown in Table 5-1. Most often, the dimensions are the same or very close to the original design, but, asking your vendor for RG-59 or RG-6 cable is like walking into a car dealer and asking for a sedan. Well, it's not a pick-up. It's not an SUV. Maybe it has four doors. What does that tell you? Not much. By the time we get to digital video, saying "RG-59" is close to meaningless.

The difference between old and new cables is, of course, technology that is more advanced. The primary technology here is plastics and especially foamed plastics. As you know, changing the plastic in the middle of a coax changes the dielectric constant and requires that a cable get smaller to stay at the same impedance. So a modern RG-6, with a gas-injected foam dielectric, is 0.275 in. in diameter, not 0.332 in. like the old solid polyethylene design.

Buying Surplus Cable

This history lesson brings us to an interesting subject. There is a lot of cable on the "surplus" market. Where it came from or how it got there is often a mystery.

Some cable comes from companies going out of business and selling their inventory. Although it might be in perfect condition, it probably is not covered by the manufacturer's warranty. See if you can find out where it came from. Most of the time, surplus sellers are very hesitant to reveal sources, fearing you will go "around" them and buy from that source.

DISCONTINUED CABLE DESIGNS

If you buy cable that has been discontinued by a manufacturer, don't expect to go back and buy more. If you do, expect to pay a lot more than your surplus sample. Most manufacturers have a minimum quantity for a "special order." This can be anywhere from

1,000 to 25,000 ft and usually depends on the complexity of the cable. If it is a 52-pair audio snake cable, with hundreds of components, the minimum might be on the smaller side. If it is a single pair, or a single coax, the minimum might be closer to 25,000 ft. If you do get a quote on 1,000 ft, you also might ask for the 10,000- or 25,000-ft price. Just like any manufactured product, the first foot costs $1,000 and the next foot costs a nickel. You might find that an additional 9,000 or 24,000 ft is just pennies different.

It is quite possible that this cable has been on a distributor's shelf for 20 years. Maybe the manufacturer has long since discontinued this design. A dusty old reel is definitely a sign for concern.

Still other cable might be a manufacturing sample, or a special construction. Often these were a "one-off" construction intended for testing and evaluation. Don't expect to find any more, if this is the case. Such a cable might have been tested or evaluated and found wanting. What were they supposed to do with that reel? The manufacturer doesn't want it, and neither does the customer who is getting ready to test and evaluate the next attempt. Hey! Let's sell it to some unsuspecting surplus customer…you!

Occasionally, cable that has design or manufacturing flaws will show up. Perhaps a multipair cable with the wrong color code was manufactured. The manufacturer might have refused to allow it to be returned or paid the distributor for the error on the understanding that the cable would be destroyed. On the way, someone figured out they could make an extra buck by selling it as surplus. So this cable finds its way back into the market, flaws and all.

There also are counterfeit versions of cables. If you want to have a certain brand name on a cable, a few manufacturers, offshore or domestic, might be willing to put whatever you wish on the cable. This might include fire ratings when no testing was ever done.

You also might check the label on the reel. Does it match what is printed on the cable? If not, the cable was respooled from one reel to another, certainly not something the manufacturer would have done. This might be a short end or the bad end of a cable from someone else's installation. The quality of the cable is completely unknown.

Of course, if you cannot trace a cable back to a manufacturer, if there is a part number on the cable you cannot follow to get details, then you are taking a risk. Just remember that you get what you pay for.

Standard Video

The list of standard video coaxes contains many of the designs from the post-war period through the 1950s. Even for analog video, one should be cautious. As you can see in Table 5-1, all the original RG-59 versions called for a copper-clad steel center conductor, indicating that the original RG-59 design was intended for high frequencies, generally 50 MHz and above.

It is at these frequencies that skin effect becomes significant so only the skin of the conductor needs to be copper. But baseband analog video is at a bandwidth of DC to 4.2 MHz. The entire channel width is only 6 MHz. This is a long way from 50 MHz. Table 1-16 shows that, at 5 MHz, a percentage of the conductor is working.

So, if you're intending to use a coax for baseband video, it should have an all-copper center. Now, you might think that all the old designs have faded away. Unfortunately, this is not true. Some of the most popular analog video coaxes, to this very day, use copper-clad steel center conductors. Many engineers who have been installing cables for decades are astonished to learn this. You might want to check the coax cable you are using.

Center Conductor

It is common to see solid and stranded center conductors. Stranded cables are generally more flexible and have better flex-life than solid center cables. Stranded cables are often preferred when used on the road, as in ENG or EFP cables.

There really isn't a problem until analog systems are changed to digital ones. Then the stranded cables won't work very well and almost all recommended designs are solid conductor. Stranded conductors can change their shape and position, so the impedance variations can be considerable, especially at high frequencies. At analog frequencies, these impedance variations are of little consequence. However, solid conductors always have better performance, which is the reason precision video cables for analog applications generally have solid center conductor.

Analog Dielectric

Polyethylene was the perfect choice for video, and 75 Ω was the perfect impedance. With the exception of its copper-clad steel center, RG-59 seemed pretty close to perfect.

There is one huge advantage of solid polyethylene: impedance tolerance. A standard analog video cable could have a maximum impedance tolerance of ± 1.5 Ω and typically ± 0.5 Ω. It has taken 40 years for foamed dielectrics to approach this impedance tolerance.

Impedance Tolerance

There is one interesting attribute about impedance tolerance. It tells you how well the cable was manufactured. The tighter to tolerance, the more consistent the product will be from batch to batch, day to day, and year to year. Eventually a cable with a very tight impedance tolerance becomes a *de facto* standard.

That impedance tolerance and the manufacturing consistency it implies really lets you know there's nothing to worry about. If you are engaged in the design or integration of an install, tight impedance tolerance means you have one less thing to worry about.

Standard Video Testing

One major difference between a standard video cable and a precision video cable is the testing done on each cable. It is rare for every foot of standard cable to be tested. More likely, one roll may be tested every so often. Most coax of this type is made in 25,000-ft "master reels;" perhaps one roll in that batch, 1,000 of 25,000 ft, will be tested.

However, there is no standard for such testing. So a manufacturer may test one roll in 25, or one roll in 100, or never! And, with less handling and testing, the price could get lower and lower. Of course, the manufacturer is assuming that what happened to the cable that was tested was exactly the same as what happened to all those rolls in between.

If it is a manufacturer with a long history and consistent quality, then occasional testing may be just fine. If this is not the case, then you are tossing the dice. Like everything else, there is no free lunch.

How to Buy Coax ... or Any Cable

In most data installations, the cable is 5 percent or less of the total cost. If we interpreted this from an audio/video perspective, it is unlikely that the cost of audio and video cable would be more than 5 to 10 percent of the total cost of an installation.

If you are spending more than 10 percent of your budget on cable, there might be something wrong. If you are spending more than 10 percent of your *time* on purchasing and evaluating your cable, you are wasting it. If you pick high-quality cable, you can give yourself a large portion of that 10 percent to use on something else.

Of course, the *time to install* the cable will most likely be the largest single time portion of an installation. To a great extent, installing cable *is* the installation. Therefore, spending an extra 5 percent, or sometimes even less, to get higher-quality cable that will be easier to install and will give you less worry and headache, will pay for itself many times over. An ounce of prevention is worth a pound of cure.

A short list for analog video cable is shown in Table 5-3.

TABLE 5-3   Analog Video Cable

Parameter	Style	Advantage/Disadvantage
Center conductor	Stranded Solid	Flexibility, flex-life, handling Performance
Dielectric	Solid polyethylene Foam polyethylene	Excellent impedance tolerance Good impedance tolerance
Shield	Single braid Double braid Foil + braid	Maximum analog coverage 95% Maximum analog coverage 97% Maximum analog coverage 95%
Jacket	Polyethylene PVC	No fire rating Standard fire rating

For plenum versions, replace the polyethylene with Teflon® or foamed Teflon and substitute different jacket compounds, such as Halar®, Solef®, or special PVC compounds that meet plenum requirements.

Analog Video Shielding

At a bandwidth of DC-6 MHz, the shield requires low resistance, so a braided shield or perhaps a French Braid® would be a good choice. These shields offer the lowest resistance to ground.

Unlike audio cables, where you can have a telescopic shield attached only at one end, video cables must have both conductors attached at both ends. If the shield is lifted at one end, that piece of equipment will seek out, electronically speaking, another ground point to complete the circuit. This is begging to have EMI and RFI enter, because there is a considerable distance that ground has to travel to get to the source or destination. Attach both ends and both conductors.

If you do not have the same ground potential at both ends of the cable, then there will be electrical flow down the cable, called a *ground loop*. The shield will feed the noise into the center conductor. How can you avoid ground loops? We'll discuss it further in Chapter 8.

The Perfect Analog Cable

The perfect analog cable would start with a solid copper center conductor. Bare copper would be preferred, although at analog video frequencies, tinned copper would be acceptable. Although larger is always better, there are diminishing returns with increased size.

Primarily, there is the overall size of the cable. Recall that impedance determines the ratio of sizes from the gage of the center conductor, the distance to the shield, and the choice of dielectric. Therefore, if you chose an RG-11 design, it would be huge, even with fancier plastics. Better to start with something a bit smaller, such as RG-59/U. So we start that cable design with a 20-AWG conductor, a reasonably large size.

Solid polyethylene as the dielectric would complete the core of the cable. The impressive impedance tolerance from such a choice has been previously discussed.

Whereas a single braid can be made with coverage up to 95 percent, a double-layer braid is even better, with 97 percent coverage, as high as you can get from a braid shield. With a double braid, the resistance is even lower, allowing even more EMI and RFI to be removed.

Over that would be a polyethylene jacket. Polyethylene is among the most rugged, scuff- and cut-resistant, UV-resistant, and water-resistant materials available. This construction made a cable that is as close to indestructible as possible. Combine that with an impedance tolerance that is second to none, and you can see why such a cable dominated the broadcast market for almost 40 years.

The King Must Die

All things have their day, and cable is no exception. The cable described above, although excellent in almost every respect, had one fatal flaw. Polyethylene burns very easily, and a cable made of polyethylene would burn very well. If you're in a mining disaster and you need a candle, it would be an excellent choice.

Such a design would not pass any common test for fire rating. Of course, one could redesign the cable with a PVC jacket or a fire-resistant core material but then it would be a different cable. The impedance tolerance numbers would increase and the ruggedness would be reduced, among other things. You may have a NEC rating, but you traded a lot to get there.

Further, when high frequencies are required, such as digital or HD television, solid polyethylene, the plastic which won the war, has become easily beatable for raw performance. And the double-braid shield is not as effective at high (digital) frequencies.

Connectors

The connector of choice for baseband video is the BNC, which stands for "Bayonet–Neil–Concelman" after the two

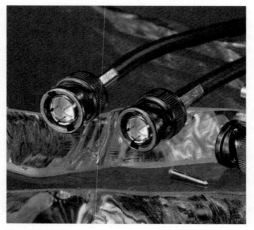

Figure 5-1   BNC connector.
(Courtesy of Trompeter Electronics, Inc.)

inventors and the fact that the connector attaches with two slots, similar to a bayonet on a rifle.

This is not to say you won't see other connectors for video, especially on older equipment. The PL-259, or UHF connector was common on early videotape machines. Table 1-13 clearly shows that the quarter wavelength at 6 MHz is 41 feet. How much difference will a connector of an inch or two make? Not much. So, even though the PL-259 has no measured impedance, it's too short to matter.

The reason the PL-259 UHF lost favor was more a matter of size. The BNC is considerably smaller. If you are wiring up a patch panel or router, with thousands of connectors, BNCs would be the only reasonable choice. The BNC, however, does not have the correct impedance either.

The BNC started life as a 50Ω connector during World War II and it remained a 50Ω connector. The wavelength of a video signal was much too long for a connector to have any influence. In fact, the BNC is intimately related to the N connector, the most common 50Ω connector for moderate-power flexible cables.

The relationship of the N and BNC can be clearly shown. Just get a male BNC and a female N and you will find that you can insert the BNC into the N. Although this is not recommended for actual operation, these will fit together and not lock, indicating a "dimensional" relationship.

One important decision is to pick one BNC and stick with it. Choosing one connector means you will need only one kind of stripper and one kind of crimper. It is a good idea to evaluate all BNCs made to fit your cable choices because standardizing with one connector for each cable type can eliminate a number of concerns.

Although it is good to have a back-up choice for emergencies, it not advisable to have two different connector types for the same cable. With two connectors, you may have to stock different parts for each. Murphy's Law says that the crimp collar for one will accidentally be dumped into the bin for the other, or some such disaster, with obvious results. Different BNCs require different stripping dimensions and different crimp tools. Standardizing with one BNC also eliminates that problem. In Chapter 9, we'll discuss just what kinds of BNCs are available, how you can differentiate among them, and which you might consider.

RCA for Video

You could put almost any connector on a coax for analog video. As long as it was a coaxial connector in style and rugged enough for the application, it would work perfectly well.

It was only logical when baseband video began to be used around the home that the tried-and-true RCA connector would be used. The only disadvantage of the RCA, compared to a BNC, is quality. The RCA is often made of thin sheet steel, stamped out, and coated with tin to prevent oxidation.

The RCA is a press-on connector and, except for some rare versions, a non-locking connector. Therefore, when inserting a connector, you never really know if it has been pushed in enough, or if the maximum contact and lowest resistance have been established. Anyone who has played around with home stereo equipment will recognize that RCA connectors have a maddening way of falling out with moderate cord pressure, just as you are returning equipment to its working position. It may be a ubiquitous connector, but it certainly is a long way from professional.

This is not to say that there aren't high-quality rugged RCA connectors with gold-plated pins and excellent strain reliefs. In fact, although most RCA connectors are solder type, you can get RCA connectors that crimp onto coax in much the same way as a BNC does. These are probably the highest-quality RCA connectors out there and approach a BNC in terms of strain relief and ruggedness.

Many times, a BNC will be attached and an adapter inserted from BNC to RCA. Although these simplify the connection, they add depth and weight to the RCA connections on the back of equipment. This added weight and depth, including that of the attached cable, makes these pieces even more likely to fall out at inconvenient times.

As stated in the S/PDIF section on digital audio, these high-quality RCA connectors, no matter how beautiful, are certainly not 75 Ω. That's fine in this case because the connectors are not long enough to make a difference.

It's the basic nature of the RCA as a non-locking connector that automatically puts it in the second tier compared to a BNC. For this reason, there are few, very expensive RCA connectors that employ some kind of locking mechanism once the plug is inserted. Therefore, it is ironic that some of these locking RCA plugs are so powerful that they can "extract" (that is, rip out) the female receptacle on the back of many types of home audio and video equipment. It just shows that this connector was never intended for the heavy-duty professional world of the BNC.

F for Video

The other connector often pressed into service is the F connector. This presents a whole different set of problems. We will discuss the F connector in its intended application, multichannel CATV/broadband delivery, in Chapter 7. The F connector most often is made of aluminum, inexpensive, reasonably rugged, and, like connectors for analog video, it doesn't matter what impedance it is.

The key problem is that the center conductor of the cable serves as the connection. On CATV/broadband coaxes, which use copper-clad steel, the center conductor is very stiff, because of the steel, and makes a reasonable center contact.

On baseband video cable, the center is bare copper or possibly tinned copper, which is a soft metal. It is annealed specifically to make it soft and flexible, which makes it a very poor choice to serve as the center contact in any connector.

If you use baseband cable in an F connector, you have to be especially careful when inserting the male connector, now with a soft copper conductor, into a female receptacle. It can easily bend or bunch up, producing less than ideal contact or even intermittent contact.

There are captured-pin F connectors where the center conductor can be inserted into the center pin. This is preferable to using a soft copper wire as the center connection. However, in some captured-pin F connectors, the center conductor is pushed into the connector, making a connection by this pressure alone. It is not crimped like a BNC. Therefore, many of the same problems of conductor bending or bunching up are not resolved. The only advantage is that inserting the male plug into the female receptacle is now as easy, rugged, and repeatable as a connector using copper-clad steel cable.

There are some captured-pin F connectors that have crimp center pins. This type of connector requires a special crimp tool, but if that is the connector on the equipment you have chosen, then that's what you'll have to use.

Precision Video

Precision video cable is different from standard video cable in a number of ways. First and foremost, it is tested differently. Most often, precision video cable has every foot tested. Table 5-4 is a list of many of the items that can be tested.

These measurements can be taken at different times in the process. For instance, gage size can be measured with a laser micrometer as the conductor is entering the extrusion machine. Likewise, the dimension of the dielectric can be measured during the extrusion process to determine physical dimensions and capacitance.

TABLE 5-4   Precision Video Cable Parameters

Parameter	Unit	To show...
Gage	AWG	Resistance of conductor
Resistance	Ohm (Ω)	Length of conductor
Capacitance	Picofarads (pF/ft)	Dimension of extruded dielectric
Impedance	Ohm (Ω)	Ratio of sizes
Impedance Tolerance	Ohm (Ω)	Variations in constituent parts
Attenuation	Decibel (dB)	Frequency response (capacitance)
Return Loss	Decibel (dB)	Variations in constituent parts

Modern extrusion processes often use the measurements made on the fly, such as those just mentioned, to control the extrusion process itself. This "feedback loop" can, for instance, change the pressure used to squeeze the melted plastic onto the wire, based on the dimensions shown by a laser micrometer reading the finished core dimension. Once the correct dimension is inserted, the machine will continue to produce cable with that dimension until the reel is full or the extruder runs out of plastic.

These extruders often run hundreds of feet per minute, so it would be impossible for any human operator to make appropriate adjustments. These adjustments are made hundreds of times each second. It is no exaggeration to say that modern coax could not be made without computerized factory machines.

Often there are dozens of interactive feedback loops. The temperature of the factory, the humidity of the air, the percentage of air in foamed plastic, the speed of the screw that creates the pressure on the melted plastic, and even things as exotic as the temperature of the conductor at various stages contribute to a precision product.

The proof, as they say, is in the pudding. In cable, the proof is in the performance after installation. It is no wonder that there are only a few manufacturers who have the expertise to produce high-quality cables. As we progress from analog video to digital and HD video, this number will get even smaller.

Analog Equalization

Analog video cable has attenuation and a natural frequency response curve (Figure 5-2). That curve is based on the capacitance and resistance of the cable. Two cables with identical components will have identical frequency response curves. Two cables with different gages or different plastic formulations in the core will have different frequency response curves.

If you send analog video signals short distances, the loss of high-frequency information is minor. If you intend on

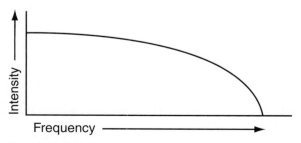

Figure 5-2   Cable frequency response.

sending analog video signals for long distances, you will need to compensate for the high-frequency loss by inserting an equalizer into the circuit.

I have heard many opinions about which distances require an equalization (EQ) circuit. Some say 100 ft, others say 1,000 ft. To be sure, some of this depends on the ability of the receiving device to recover the signal. So a quick look at the owner's manual for such a device would be in order.

If you use distribution amplifiers, intended to drive signals long distances, they often come with an EQ option.

What does equalization do? It makes up for the natural loss in a cable. Figure 5-2 shows a frequency response curve for a piece of cable and Figure 5-3 shows the response of the EQ card for that cable (dotted line).

The thick line in Figure 5-4 represents the resultant curve of the EQ card for the cable in question. In an ideal world, the results would be flat frequency response, as shown.

The curve on the EQ card is pre-set by components on the card, and the actual curve of the cable is determined by its resistance, capacitance, and the length of the cable. Some

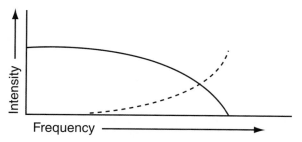

Figure 5-3   Equalizer response.

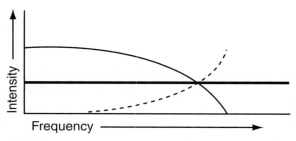

Figure 5-4   Resultant curve.

EQ cards are fixed for a specific distance, whereas others allow you to control the "slope."

A Two-Cable Installation

The only real problem with EQ occurs when there is a second cable type in an installation. Unless all the cables are short, you may require two different EQ cards.

Murphy's Law says that someone will put an EQ card for the first cable into a slot connected to the second cable. This usually occurs when there is existing cable, and you want to install some newer cable on top of it. Most often, the old cable is a solid dielectric type, and the new cable is a foamed dielectric. Each of these cables has a different attenuation curve. Figure 5-5 shows the same cable as Figure 5-4, with the frequency response of a newer foamed-dielectric cable.

Figure 5-5 shows the result of using the new cable with the old equalizer.

The resultant curve, the black line between the curves, is no longer straight. It shows a rising frequency response, which means that the high-frequency components of the sig-

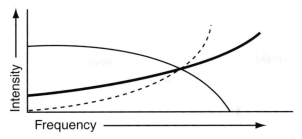

Figure 5-5 Solid dielectric cable frequency response.

nal are rising in level. This can overload the inputs of destination devices and produce distortion.

The reverse can also occur, as shown in Figure 5-6. This is where the old cable is accidentally attached to an equalizer intended for the new cable. In this case, the result will be a dropping, not flat, response dramatically reducing effective distance.

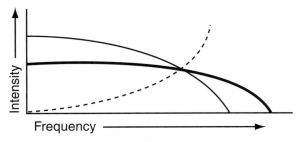

Figure 5-6 Foam dielectric cable frequency response.

One-Cable Installation

The solution is to standardize as much as possible on one cable because you would need only one EQ card, one type of BNC, and all the tools necessary could be adjusted for that one cable. This often requires the removal of old cable before the installation of new cable. This can have some added advantages that we will discuss in Chapter 8. The simplest one-cable installation is new construction in a new building.

Composite/Component

Video contains a number of discrete signals. These can be sent in two different ways, as a combined as a "composite" signal. Most video interconnections are made with composite video.

However, if greater detail is desired, the video signal can be divided into its component parts. The most common analog version is RGB.

RGB

Component signals are carried on cables that contains multiple coaxes. Each coax in intended to carry a different component part. Table 5-5 lists a number of components, the coaxes required to carry them, and the color codes of the resultant cable.

With any component system, it is necessary to carry a synchronization signal, or "sync." This allows the red, green, and blue signals to be realigned.

If only three coaxes are used, the sync signal is included on the green coax. For even greater accuracy, the sync (S) can be given its own fourth coax. Such four-coax cable is often designated RGBS. For even greater accuracy, sync can be divided into horizontal (H) and vertical (V) sync, for a five-coax cable, RGBHV.

TABLE 5-5   RGB Components

Number of Coaxes	Component Part	Color Code
1	Red component	Red
2	Green component	Green
3	Blue component	Blue
4	Sync/horizontal sync	No standard (white?)
5	Vertical sync	No standard (yellow?)

There are other ways to deliver component signals. Some are subtractive, such as Y, R–Y, B–Y, or C–Cr–Cb. The nature of the signal may be different, but the delivery is identical to that of RGB and the same cable is used for any of these applications. RGB sometimes is referred to as "GBR."

Timing

The signals on the three coaxes are critical in terms of arrival. All coaxes have delay (page 46). The key to RGB cable is that the delay, or "timing," of the three component signals is close enough to produce an adequate image.

If the signals are too different in arrival time they may be too far out of time. The maximum allowed time differential is 40 nanoseconds (nsec). At that point, the arrival times of the three component signals will be so great that details will be lost because the signals don't line up. More than a 40-nsec timing error is considered "non-broadcast quality."

Multicoax cable intended for use with RGB signals should have a specification for the maximum expected time error. If no timing value is available, then you will have to time the cable yourself. This is a lengthy process. Using unbundled coaxes or loose single coaxes also requires hand timing. We will discuss timing by hand in Chapter 9.

Timing variations occur in RGB cable because the three or more coaxes are not and cannot be exactly the same. They will have slightly different parts, resulting in slightly different velocities, slightly different impedances, or other differences. These can translate to a difference in the time it takes the signal to travel along a given length of cable.

The key is that the physical length of the cable and the electrical length will be two different distances. What you really want is for the two to be identical.

If the physical and electrical lengths were identical, you could just cut the cable at the desired length, put on the connectors, and attach it. That is exactly what pre-timed cables are supposed to do. Therefore, you need to know the "maximum timing variation" specific for these cables, and that variation needs to have a distance attached.

HORROR STORY #1,000,001

A major installer was working on a project that required lots of RGB monitoring. He completed the job, which was in the Midwest, and returned home on Friday. The following Monday he got a phone call from the customer: "Everything is out of time." He thought this was impossible, so he sent someone to check it out. Sure enough, every RGB was way out of time. He sent a crew that retimed and reconnected everything. The next Monday, the customer called again: "Everything's out of time again." After the fourth round, he finally realized what was happening. The RGB cable used a soft foam insulation. Because the cable was bent, the center conductor slowly began to migrate through the foam, and each coax would migrate a little differently. This changed the impedance and capacitance of the cable and affected the timing by slowly moving it out of the 40-nsec window. After a few weeks, the migration finally stopped and one more round of timing and connecting finally finished the job. I am sure that job had a negative profit margin!

TABLE 5-6   Timing Variations

Maximum Timing Variation	Distance to 40 nsec Maximum
1 nsec/100 ft	4000 ft
2 nsec/100 ft	2000 ft
3 nsec/100 ft	1333 ft
4 nsec/100 ft	1000 ft
5 nsec/100 ft	800 ft
6 nsec/100 ft	667 ft
7 nsec/100 ft	571 ft
8 nsec/100 ft	500 ft
9 nsec/100 ft	444 ft
10 nsec/100 ft	400 ft

Table 5-6 shows timing variations, and how far a signal can go on a cable until it reaches the 40-nsec maximum timing error. The maximum timing variation is between all coaxes in that assembly, or the farthest apart in time any two coaxes will be in the assembly.

You also should be aware that there are maximum timing numbers that are most desirable. Then there are average typical, and nominal timing numbers that are a bit more vague. Always get maximum numbers if possible.

A manufacturer may list a very short distance, shorter than the 100 ft in Table 5-6, and come up with a very small timing difference. To generate the numbers in the second column, divide the nanoseconds in the left column into 40 (maximum timing variation of 40 nsec) and then multiply by the footage specified in the left column. If the footages are short to show extra low timing variations, doing the distance calculation will let you know the truth.

Because most RGB coaxes are made of groups of coax cables, the internal coaxes are often quite small to keep the overall size down. Very small component cables have two problems. First, it is hard to make coaxes that are precise and close to identical. Therefore, most miniature RGB cables are made into short assemblies, only a few feet long. Sometimes these tiny coaxes are not even 75 Ω. These assembled cables are too short to make a difference, no matter how poor the coaxes.

The second problem with small coaxes is basic attenuation. These tiny coaxes use equally tiny parts, especially tiny center conductors. Therefore, these cables will have high attenuation and lose signal strength rapidly. Longer assemblies of miniature coaxes become unusable because of signal loss long before the signal reaches the maximum timing error in Table 5-6.

Pre-timed RGB Cable

The whole purpose of timing an RGB cable is to get within the 40-nsec window and even less, if possible. Therefore, time and labor can be saved if pre-timed cable is specified. Pre-timing is done by the manufacturer of the cable. It requires very precise cable manufacturing to pre-time the cables. The easiest way is to use internal cable that is one color. Then you can run that internal cable as a single batch. Because the three, four, or five cables are made from the same cable, they will be very closely matched in basic parameters.

Today's RGB cables are color coded. The individual jackets are red, green, and blue, and possibly two other colors. This means that the five cables will be done in five batches, at different times, by different operators, maybe even on different machines. Making electrically identical cable is no mean feat.

It is no wonder that many RGB cables, especially miniature versions, are limited to short distances of less than 100 ft. The problem isn't just small-wire resistance and attenuation, it's also timing errors that were inherent to the design and manufacture of that cable.

Table 5-7 shows the timing differences between coaxes based on velocity. Of course, there is basic resistance and capacitance, read as attenuation, and impedance variations, shown as return loss. Velocity is the only parameter compared in this table.

TABLE 5-7   Comparing Velocities

First Coax	Second Coax	Timing Variation/100 ft
83%	83%	0 nsec
83%	82%	1.47 nsec
83%	81%	2.97 nsec
83%	80%	4.52 nsec
83%	79%	6.1 nsec
83%	78%	7.72 nsec
83%	77%	8.39 nsec
83%	76%	11.09 nsec
83%	75%	12.85 nsec
83%	74%	14.35 nsec

However, the distances show in Table 5-7 often are not achievable because of the basic attenuation of the cable. Certainly distances of 1,000 ft or above probably are not achievable on standard RGB cables.

It is for this reason that larger multicoax cables have been introduced. These go up to multiple RG-6 coaxes, with the RGB color code. These "extra large" component cables have very low loss and can easily exceed the 1,000-ft barrier. These component cables are large, fairly stiff, and expensive.

This brings us to an interesting point. Cables with better impedance tolerance are easier to time. Their impedance tolerance gives you a hint of how close their physical length will be to their electrical length. So, ironically, the old solid polyethylene cables are much easier to time and assemble that the newer foamed cables. This will be a significant factor when we discuss installing RGB facilities in Chapter 9.

Multicoax Beyond Six

You also can find multiple coax cables up to ten or even twelve coaxes. These are popular for applications where more than one RGB signal is being fed, such as multiple projectors in command centers, rock concerts, or political conventions.

Multiple coax cables also can be used to feed component and composite or even analog and digital signals at the same time. Often RGB projectors are backed up with composite versions of the RGB signal. If the RGB fails, meaning that one of the three colors fail, then the operator can switch to the composite input of a projector. It won't have the same detail as an RGB image, but at least there will be a picture on the screen!

We'll explore other uses for multiple coaxes in Chapter 9.

S-video

Super-video (S-video) is also called S-VHS® and Super-VHS®. The Super-Video Home System was originally intended for high-end home installations, where large projection televisions or other large-screen images would be used. S-video splits the video signal into two parts, the black-and-white portion (luminance) and the color portion (chrominance). It is also called Y–C, where Y stands for luminance and C for chrominance.

The simplicity of this system has made it very attractive for broadcasters on a budget. S-video cameras, even the top-of-the-line versions, cost a small fraction of their full-throttle broadcast cousins. For less critical video, such as industrial, training, and corporate or short-lived content, such as news footage, S-video can be a very cost-effective alternative.

The limitation to S-video is distance. It has a much lower voltage than the 1-volt of professional video, and the cable used, a dual video cable, is very small. The reason is the connector, a DIN plug. Even the DIN with the largest opening can barely fit most S-video cables.

S-video cables come in a number of sizes. Table 5-8 lists three common sizes.

TABLE 5-8   S-Video Cables

Style	Dimensions (in)	Center Gage
Dual figure-8	0.110 × 0.230	30 AWG
Plenum dual	0.107 × 0.214	30 AWG
Round	0.255	30 AWG

The distance is limited to approximately 50 ft. Converting to cables with larger conductors can double this distance. You can purchase amplifiers intended to send S-video considerably longer, but they are an added expense. The original intention of S-video was to send high-quality video from device to device within a single room. The small size of the center conductors in this cable attests to this short-run design.

Patch Panels

Patch panels for analog video are significantly different than analog audio patch panels. Of course, we are dealing with unbalanced coax. Therefore, patch cords and patch panels are also unbalanced and coaxial.

These components are reasonably close to 75 Ω, although it would take a lot of jacks and plugs to get up to the 41 ft of a quarter wavelength at 6 MHz, the channel width of a video signal.

In truth, many patch jacks and plugs are not very near 75 Ω, and certainly by the time we talk about digital video, they are significantly far from that impedance. Analog video is very forgiving.

We will discuss the specifics about patching analog video in Chapter 8.

Video on Twisted Pairs

One option that is surprising more than a few old video engineers is using twisted pairs to run video. In fact, this is not new at all. Systems were designed to run video on "twinax" computer cable in the 1970s. The noise rejection of twisted pairs and the belief that it was harder to tap into such a system were the motivation factors.

Twinax was the first attempt at an impedance-specific twisted pair. Although nowhere near a coax in impedance tolerance, twinax was significantly better than previous twisted pairs. With the rise of category cables, especially the emerging Category 6, and the bonded pairs of some of those cable designs, interest was revived on running video on these cables.

Table 5-9 shows the required performance for analog video and what can be expected from Category 5, 5e, and 6 cables. The demands of analog video have proven to be one of the hardest things to transfer to UTP and the crosstalk requirements in Table 5-9 show why. Also, there is the required conversion from unbalanced to balanced. This can be accomplished by using a balun. Next is the conversion of impedance from 75 to 100 Ω. This too can be accomplished with a balun.

TABLE 5-9   UTP and Analog Video

	Value	Category 5	Category 5e	Category 6
Format	Unbalanced	Balanced	Balanced	Balanced
Capacitance (pF/ft)	20	15	15	15
Impedance (Ω)	75	100	100	100
Gage	Varies	24 AWG	24 AWG	24 AWG
Shield	Yes	No	No	No
Crosstalk (dB) at 100 m (328 ft)	−60 at 4.2 MHz	−53 at 4 MHz	−56 at 4 MHz	−61 at 4 MHz

Figures 5-7 and 5-8 show some of the baluns available to convert analog video to UTP. These range from RCA input for consumer applications to BNC input for professional applications.

However, there are three basic limitations that cannot be solved by a balun. First is the gage of UTP, which is 24 AWG. Some bonded Category 6 cables actually use the largest size wire that will fit into an RJ-45 modular connector, about 23.5 AWG. This difference in resistance and surface area at 100 MHz adds about 15 percent to the effective distance. At analog video, it is a minor factor. Coax has no such limita-

Figure 5-7 RCA video balun.
(Courtesy of ETS, Inc. www.etslan.com.)

Figure 5-8 Professional video balun.
(Courtesy of ETS, Inc. www.etslan.com.)

tion and increased distance can be had by simply installing a larger coax.

THE LONGEST ANALOG RUN ON UTP

I once saw a demonstration of video on a bonded-pair Category 6 cable. A balun was attached to a broadcast-quality camera and the signal was fed down UTP to the control room on the other side of the building, where it was converted by another balun and went through a router and switcher to unity gain (that is, no amplification). It went in to another balun and back to our room, where the signal was fed into a video projector. This was an auditorium, with a video projector and the video image was about 30 ft across. Most people thought the image was being fed from the camera directly into the projector and they had to come up and unplug the UTP to prove otherwise to themselves. Although this analog video signal was not analyzed on a vectorscope or waveform monitor, it was excellent. The total distance was 1,400 ft.

The second factor that cannot be helped by baluns is the impedance and especially the impedance tolerance of the cable. Coax is the king of impedance tolerances. Modern digital video cables are rated a maximum of ± 1.5 Ω and typically ± 0.75 Ω. The best UTP, bonded-pair Category 6, has a maximum impedance tolerance of ± 10 Ω, typically ± 7 Ω.

The third concern is crosstalk. You can see from Table 5-9 that the requirement for crosstalk is -60 dB for analog video. It is a *de facto* standard. If you believe it should be more or less than this value, you can make your judgment appropriately.

Generic Category 5, at only 4 MHz, has a crosstalk number of only −53 dB, which is adequate if this is a surveillance camera. It is also adequate for most consumer/home theater applications. Most television received in the home rarely exceeds 50 dB of signal-to-noise ratio and is often closer to 40 dB. DVDs, from their analog outputs, can exceed a signal-to-noise ratio of 60 dB.

By the time you are at Category 5e, this ratio has improved to 56 dB. Although it is better, it does not meet our 60-dB requirement, and the same arguments apply. Only the bonded Category 6, at 61 dB, passes that magic 60-dB mark. These numbers apply to 100 m (328 ft.) of category cable. Longer distances produce more crosstalk.

Some may take me to task for using *crosstalk* and *signal-to-noise ratio* interchangeably. In the data world, I might be more correct to use *attenuation-crosstalk ratio* (ACR) to represent *signal-to-noise ratio*. There is an even harder version of ACR called "power-sum ACR" (PSACR), where three of the pairs are driven and the resultant signal is read on the fourth pair. All pair combinations are measured and averaged to show the data in Table 5-10. which also shows the PSACR for the cables shown at 4 MHz.

TABLE 5-10   Signal-to-Noise on Category Cable

Category	PSACR (dB)
5	49.2
5e	49.2
6	59.5

If you agree that a 60-dB signal-to-noise ratio is required, then theoretically, you cannot run four video signals simultaneously. If 59.5 dB is sufficient and 60 dB is only a suggested value, then do as you wish. Again, this is 100 m (328 ft).

Other UTP Video Baluns

Figures 5-9 and 5-10 show examples of baluns for S-video and for RGB.

S-video, with its limitations of gage size and system voltage, is greatly helped by such baluns. Even though these are passive boxes, distance is extended to 500 ft or more for the S-video signal. The audio pairs are extended well beyond 1,000 ft.

You might think that RGB, with its timing requirements, would be difficult for twisted pairs. In fact, UTP does very well in this regard. The maximum timing variation, called *skew/delay* in the premise/data world, is 45 nsec per 100 m (328 ft). This is amazingly close to the 40 nsec for broadcast quality RGB.

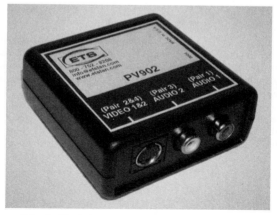

Figure 5-9   S-video balun.
(Courtesy of ETS, Inc. www.etslan.com.)

Figure 5-10   RGB balun.
(Courtesy of ETS, Inc. www.etslan.com.)

Bonded-pair Category 6 does even better, with a maximum delay of 18 nsec per 100 m, or an amazing 5.48 nsec per 100 ft. This style of twisted pairs rivals the best timing on multiple coaxes, with the best coaxes running a maximum of 4 nsec per 100 ft.

This is the reason there are many installations with projectors in conference rooms wired up with Category 5, 5e, or 6 cable. The cost of the balun at each end is offset by the simplicity of the installation and the reasonable price of the cable.

6

Digital Video

Data Rates and Standards

Digital video standards are overseen by the Society of Motion Picture and Television Engineers (SMPTE). Digital video is a sampled version of analog video. There are many data rates that apply to digital video. They cover the various versions of analog video. Table 6-1 lists the basic data rates and the actual occupied bandwidth of each.

TABLE 6-1   Digital Data Rates

Data Rate	Clock Frequency (MHz)	Comments
143 Mbps	71.5	Converted standard NTSC
177 Mbps	88.5	Converted European PAL
270 Mbps	135	Converted NTSC or PAL RGB component
360 Mbps	180	Widescreen composite
540 Mbps	270	Compressed widescreen component
1.485 Gbps	750	Uncompressed widescreen component

There are composite and component versions of domestic and European standards. There are different screen sizes, different frame rates, and different ways of reproducing the image, such as interlace or progressive scanning. Each has an impact on the final occupied bandwidth of each digital video system.

We will discuss each of these standards, not how the digital image is created, but how to ship it from place to place with minimum loss.

Composite and Component

These choices for digital video are the same as for an NTSC signal or for an RGB signal. Composite signals carry all the picture information in one bitstream. Component signals carry the "components" of the picture, the RGB portions, as separate signals or "words." These are digital words run serially, one right after the other, without affecting the timing. There is a buffer in the receiving device that divides, organizes, and times the information so that the final analog output is perfectly in time.

Therefore, composite and component digital run on single coaxes. You don't need separate coaxes to run digital component signals, just more bandwidth. Table 6-1 shows that a composite NTSC digital bitstream is 143 Mbps, and the component version of the same signal is 270 Mbps, almost double the data rate.

Screen Sizes

There are two screen sizes in the SMPTE standard. These are expressed by the ratio of width to height. One can buy televisions or monitors that are tiny to huge, with the same size ratio.

The standard television and its digital counterpart have a ratio of 4 to 3. In the motion picture world, this is expressed at a ratio of 1.33 to 1. The second ratio, widescreen, is 16 to 9, or a ratio of 1.77 to 1. The latter ratio does not correspond precisely to any motion picture image.

Image Type

There are also two video image types expressed in the standard. The first is the *interlace scan*, the standard for pictures on a consumer television. The screen is drawn with every other line in one-sixtieth of a second. After the odd-numbered lines are drawn, the image in drawn with the even-numbered lines in the next sixtieth of a second. Thus, in one-thirtieth of a second ($2 \times 1/60$), an entire frame, or picture, is drawn. This is then a 60-field, 30-frame per-second system. Only half of a picture appears at any one instant. Due to an effect in our eyes and brains called *persistence of vision*, we integrate the two half-images into one.

The second image type is the *progressive scan*, where the lines are drawn one right after the other until the image or frame is complete. This provides much greater resolution and detail and used most often for computer monitors and other graphic displays. The bandwidth of progressive systems is greater than that of interlaced systems.

The type of display does not affect the choice of cable except as a function of bandwidth. The higher the bandwidth, the more critical the cable choice is.

Digital Cable Choices

Because the bandwidth of the digital video signal is significantly greater than that of an analog video signal, a number of factors that were minor in analog become important in digital.

Although standard video cables can be used for digital signals, distances will be limited because the high frequency, essential for good digital performance, is compromised. Further, these standard cables are not constructed or measured for performance at these high frequencies.

Everything that was important to analog cable is important for digital applications. But the order of importance has changed. Impedance, for instance, which is moderately important to analog, is very important, sometimes critical, for digital and high-definition installations. That makes impedance tolerance more important. So the ultimate expression of impedance variation, return loss, is also important.

Return loss shows impedance variations in cable, connectors, and associated hardware. Figure 6-1 shows the return loss of a cable. Because return loss shows signal reflections, or a returned signal, a lower number is better. That is the reason you see spikes, sometimes called *grass*.

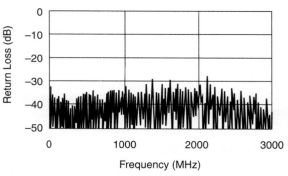

Figure 6-1   Return loss.

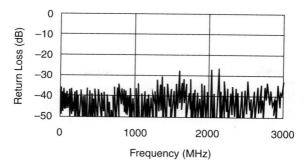

Figure 6-2 The next 100 ft.

This particular cable is a gas-injected foam precision video cable in an RG-59 size. This cable is generally –30 dB or better, which is excellent performance. You will also note a spike or two above the –30 dB line. Where do these spikes come from?

Figure 6-2 shows the next 100 ft of the same reel. Its return loss is different. Why would that be? It has to do with the nature of return loss.

What Causes Return Loss?

Return loss can be caused by any one of the factors listed in Table 6-2. All of these factors eventually affect the impedance of the cable. These impedance variations cause the signal to reflect back to the source.

TABLE 6-2   What Causes Return Loss?

Part	Attribute	Effect
Center conductor	Wrong size	Wrong impedance
Center conductor	Varies in size	Varied impedance
Dielectric	Wrong dimensions	Wrong impedance
Dielectric	Varies in size	Varied impedance
Dielectric	Incorrect foaming	Wrong impedance
Dielectric	Variable foaming	Varied Impedance
Dielectric	Foam too soft	Bending changes impedance
Shield	Braid too tight	Wrong impedance
Shield	Varies in tightness	Varied impedance
Jacket	Print wheel pressure	Affects impedance

So then, what are the spikes in Figures 6-1 and 6-2? They could be any of the factors in Table 6-2 or even some more exotic manufacturing problems, such as a gear out of round or a flat ball bearing. In other words, we have no idea what causes these spikes.

One thing is certain. Every time an extruder, braider, or other machine used to make this cable is torn apart, and all the gears, wheels, and bearings are replaced, the machine will make wonderful cable. Top-of-the-line manufacturers do such overhauls on a regular basis. Such an overhaul on an extruder costs more than $100,000, and this is done often without any specific reason or requirement. The cable coming off this machine may be completely acceptable. After an overhaul, the cable made on that machine is amazing.

Structural Return Loss

When researching cables and comparing manufacturer's specifications, you might run across something called *structural return loss*. This is not the same as return loss.

Structural return loss is commonly used in the broadband/CATV industry. It is a way of looking at the large impedance errors in a cable. To measure structural return loss, a coax is attached to an analyzer. The analyzer is then matched to the impedance of the cable. The impedance variations are then read and displayed.

This seems reasonable, but the key here is that the analyzer is matched to the cable. Therefore, if the cable is 73, or 68, or 79 Ω, that error is "nulled" and only the variations from that impedance are read. This might be a reasonable test for broadband cable, but it's not a good test for digital video cable.

For one thing, the end user cannot "adjust" his equipment to match any impedance variation in the cable. It's set up for 75 Ω, so everything had better be as close to 75 Ω as possible. Any variation around 75 Ω will simply add to the return loss and limit the effective distance the cable can run.

Therefore, if you are asking a manufacturer for return loss data and presented with structural return loss, politely hand it back and ask for return loss data. While you're at it, ask if those numbers are maximum numbers (that is, guaranteed return loss) or typical or nominal numbers that indicate what can be commonly expected from the cable. Of course, guaranteed numbers are always your best bet.

Crushing a Cable

Cables are deformed when they are installed. A very gentle and careful installation won't deform cable. If it is done by a couple of out-of-work wrestlers, the effect might be a little different. How much reduction might you expect from a deformed cable?

PUT THIS CABLE IN AND STEP ON IT!

I had the chance to step on the ball of cable that resulted from this testing. It was still attached to a network
analyzer showing return loss. One interesting fact was
that, when I stepped on the cable, I could clearly see
the effect in the next sweep. When I took my foot off,
some of the effects reversed, whereas others
remained. The effects that reversed were probably due
to the use of a high-density hard-cell foam dielectric.

Figure 6-3 shows the next 100-ft section of precision digital RG-59 cable. A number of people twisted it, folded it, tied
it into knots, and stepped on it, until it was a ball of cable.
It is a testament to the quality of this cable that its return
loss rose "only" about 10 dB. Other similar cables might not
have fared as well.

This cable was still good because of crush resistance, the
ability to withstand deformation by using high-density

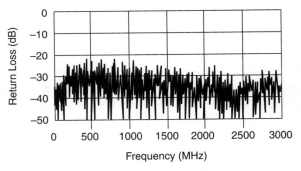

Figure 6-3 Crushing a cable.

hard-cell foam in the dielectric. Most nitrogen-gas–injected foam is very soft, so it is easily deformed, which leads to impedance variations and ultimately poor return loss.

Crush Testing

This type of testing is new to digital video, but it is very informative. It is based on testing done for the broadband/CATV industry. The test consists of attaching a sample of cable to a TDR. This measures the impedance along the length of the cable.

A 4-inch square steel plate is placed on a sample of cable. Pressure is applied to the plate that squeezes the cable at the rate of 0.2 in. per minute. When a 3 Ω change in impedance is noted, the pressure applied is read off the machine.

Tables 6-3 and 6-4 show some comparative results.

TABLE 6-3   RG-59 Crush Testing

Precision Digital RG-59	Average Crush Resistance
Manufacturer X	275 lb
Manufacturer Y	164 lb

TABLE 6-4   RG-6 Crush Testing

Precision Digital RG-6	Average Crush Resistance
Manufacturer X	405 lb
Manufacturer Y	206 lb
Manufacturer Z	213 lb

Both tables clearly show that not all cable is alike. Those designs that use soft foam are easily crushed and have almost half the crush resistance of other cables. The lesson is to get this information from a manufacturer or distributor so that you can compare cables.

It is also interesting to note that the larger cable (RG-6) has more crush resistance than the smaller cable (RG-59). This is mostly a factor of size because it takes a greater deformation in a larger cable to create a 3 Ω change. Larger cables, such as precision digital RG-11, have even more crush resistance.

Precision Digital Cables

Cables for digital applications come in a number of sizes. Table 6-5 shows a number of common types and their advantages and disadvantages. Each of these cables has advantages when used in specific applications.

TABLE 6-5   Precision Digital Cables

Cable Type	Outer Diameter (in)	Center	Advantages	Disadvantages
Miniature	0.150	Stranded	Flexible	Higher return loss
Miniature	0.159	Solid	Small	Shorter distance
RG-59	0.235	Solid	Universal size	Moderate distance
RG-6	0.275	Solid	Size versus performance	Stiff, slightly large
RG-7	0.320	Solid	More distance	Large, stiff
RG-11	0.405	Solid	Longest distance	Very large, very stiff

Miniature cable

The miniature cables are intended primarily for truck cabling, where size and weight are major concerns. The solid conductor is preferable for digital and HD applications. This cable also can be used where it must be hidden on sets or stages, or where weight is a factor, such as cables that must be carried repeatedly by hand.

There are also stranded miniature cables. Although these are more flexible than solid conductors, the impedance variations and resulting return loss make this style a lesser alternate to the solid-center version. It is also slightly harder to put connectors on a stranded center. Besides, a small cable is naturally flexible based on its size. Thus, solid-center miniature cable is easy to install.

RG-59

The RG-59 digital cable is the most popular size in the world. Its reasonable size allows it to be used in analog or digital applications with a wide range of connectors. It is the most common digital cable used with RCA connectors for S/PDIF applications, for instance. In fact, most precision video cable is not really RG-59. It is the right diameter but the center conductor is 20 AWG, much larger than standard analog RG-59 video cables that are 22 or 23 AWG. It is the gas-injected foam dielectric that allows the center conductor to be 20 AWG because the dielectric constant is so much lower than solid polyethylene.

RG-6

The RG-6 digital cable is the most popular cable for HD television upgrades. It seems to be a good compromise between size, stiffness, and performance. More than 80 percent of all digital/HD television station upgrades have been done, or will be done, with this size cable. Interestingly, it is smaller and less expensive than the most common precision analog video cable.

Stiffness versus Flexibility

Speaking of stiffness, it should be apparent that stiffness can be an indication of performance quality. Any broadcast engineer will tell you that the ultimate antenna cable (transmission line) is hard-line. That is a solid copper pipe!

Sorry to say, video designers and installers are now in the same boat. A solid copper pipe would also be the ultimate HD video cable. The more flexible the cable, the more that flexibility may have come from soft foam, which may lead to impedance variations and return loss. That is not to say that high-quality, high-performance digital cables are totally inflexible. They have significantly more "memory" than stranded cables. "Memory" is recognized by how a cable stays in a certain shape after being bent by hand. It retains the bent shape. Flexible cables have very little or no memory. All bends straighten when you let go.

Memory can be a good thing. If you are installing cables, a little memory keeps cables neat. It makes them easier to dress. And, like our discussion of flexible snake cables, cables that are on the stiff side are often easier to install in conduit than flexible cables. They will not readily grab the side of the conduit as the cable is pulled through.

Whereas ultra-flexible cables are appropriate for analog audio, and acceptable for digital audio, such flexibility for digital video would be a serious compromise in performance. The key question to ask a manufacturer of such cables is, What is the guaranteed return loss of these cables to at least 2.25 GHz? (We'll talk about that particular frequency on page 261). I bet that few of these manufacturers have ever measured return loss or they measured it a long time ago at frequencies that are now too low to matter.

Distance

The distance that digital cables can go is based on a formula used by the SMPTE. The formula, shown in Figure 6-4 is for SMPTE standard 259M and includes all NTSC and PAL formats up to 360 Mbps. The formula in Figure 6-5, is for SMPTE standard 292M and applies to uncompressed HD television applications.

Maximum length = 30 dB loss at $\frac{1}{2}$ the clock frequency

Figure 6-4 Formula: serial digital distance.

Maximum length = 20 dB loss at $\frac{1}{2}$ the clock frequency

Figure 6-5 Formula: high definition distance.

Table 6-6 shows the distances when actual distance values are calculated for cables.

> **To Dream Afar**
>
> Just imagine! If that digital RG-11 will go 2,730 ft at 71.5 MHz, imagine what it could do at 4.2 MHz analog video (with the correct equalization, of course).

The HD formula significantly restricts the distance of any cable. So how accurate are these numbers? They were cho-

TABLE 6-6 Digital Cable Distance (in Feet)

| Precision Digital Cable Type | SMPTE 259M | | | | SMPTE 344M | SMPTE 292M |
	143 Mbps 71.5 MHz	177 Mbps 88.5 MHz	270 Mbps 135 MHz	360 Bpps 180MHz	540 Mbps 270 MHz	1.485 Gbps 750 MHz
Miniature	1,000	910	750	650	530	210
RG-59	1,430	1,320	1,110	960	790	300
RG-6	1,760	1,620	1,360	1,180	970	370
RG-7	2,220	2,000	1,670	1,460	1,210	470
RG-11	2,730	2,460	2,000	1,740	1,430	540

sen specifically by the SMPTE committee to keep designers and installers safe. By safe, I mean away from the digital cliff. We mentioned the cliff effect in the discussion about digital audio. It becomes more significant for digital video.

The Digital Cliff, Part 2

The same concern about digital audio and the digital cliff effect applies to digital video. The digital cliff, you may recall, occurs where bit errors rise and, in a short span of cable, overwhelm the receiving device. With its higher data rates and higher bandwidths, the digital video cliff can become even more abrupt and prominent. Bit errors occur in all digital systems. Eventually, the receiving device becomes unable to decode the data and shuts down. Where the effect is due to cable length and, hence, signal attenuation, return loss, or noise, the distance can be 50 ft or less.

How Accurate Are These Distances?

The distances in Table 6-6 are supposed to keep you safely away from the digital cliff. So where is the cliff? It's approximately double the distance shown in the table. The problem is that, with some equipment, the distance shown might be closer to correct, whereas other equipment, with newer chips and better decoding capability, could go fully double the distance with no increase in bit errors. So how do you know? You don't. If you decide to go farther than the recommended distance, you are on your own. That simply means you need to be able to read bit errors. If you have a cable listed at 300 ft in the chart and you have a device 400 ft away, run the cable and look at the bit errors. If there aren't any, you're home free. There are

> a couple of installations that have run precision digital RG-11 (540 ft in the table) to almost 900 ft with no problems. They just read the bit errors.

Of course, the problem with the cliff is that you don't know it's coming. Video engineers rarely monitor bit errors, and the test equipment to measure bit errors can be expensive. Therefore, the bit error rate can be very close to maximum on one cable and a simple addition, such as a longer patch cord or one poor choice connector, can push the error rate over the cliff.

This is the reason the distances in Table 6-6 are recommended, not guaranteed, distances. It is quite possible, with the wrong equipment, bad connectors, old patch panels, and a dozen other possible poor choices, that even these distances would not be achievable.

The real problem is how to recognize a bit error problem. Let's say you have a large digital installation and one cable is not working. Table 6-7 shows many of the possible reasons it isn't working, the last being the simple fact that the cable is too long and over the digital cliff.

Of course, a connector "incorrectly installed" deserves its own table. This list also doesn't begin to address the settings and adjustments of the machines themselves. The days when you could take a volt-ohmmeter and fix a piece of malfunctioning equipment are long gone. The complexities of a modern digital install can often confuse and frustrate even the most savvy designer or installer. The actual problem could be as simple as not plugging something into the power strip.

If your cable is not working, which item in Table 6-7 is the culprit? Of course, you have no idea, and you can come up with another dozen items for your own expanded list. The only way you can avoid such a situation is by training, education, and establishing a standard for installation procedure. That is exactly what Chapter 9 is about.

TABLE 6-7   What Causes Bit Errors?

Part	Problem	Effect
Center conductor	Wrong size	Wrong impedance
Connector	Incorrectly installed	No signal
Connector	Wrong impedance	Return loss
Connector	Unstable impedance at high frequency	Return loss
Patch panel	Designed for analog	Return loss
Patch panel	Unstable impedance at high frequency	Return loss
Cable	Incorrectly prepared	No signal
Cable	Designed for analog	Periodicity, return loss
Cable	Poor quality control	Return loss
Cable	Too long	Excessive bit errors

Connectors

There is really only one connector appropriate for digital video, and that is the BNC. Digital BNCs are different from the BNCs previously discussed. A digital BNC is specifically designed for digital and HD applications. The difficulty is shown in Table 6-8. We have gone from low-frequency analog to HD signals in only a few years.

Connector manufacturers have made a valiant effort to keep up with the break-neck speed of new technology. Unfortunately, it means that there are BNCs on the market with varying quality and consistency. Some were designed for analog applications and may not be 75 Ω. Still other BNCs may have been designed for 270 Mbps/135 MHz. There is also the true BNC, designed for HD applications.

TABLE 6-8   Video Signals, Clock, and Wavelength

Data Rate	Clock	Wavelength	Quarter Wavelength
Analog	4.2 MHz	234 ft	59 ft
143 Mbps	71.5 MHz	14 ft	3 ft
177 Mbps	88.5 MHz	11 ft	3 ft
270 Mbps	135 MHz	7 ft	2 ft
360 Mbps	180 MHz	6 ft	1 ft
540 Mbps	270 MHz	4 ft	1 ft
1.485 Gbps	750 MHz	1 ft	4 in

How are they different and how can you tell them apart? The most important way is how they are measured, and what the results of those measurements are.

The third harmonic

With any device intended to carry digital signals, it is essential that it be able to carry a bandwidth in excess of the actual clock frequency or data rate. As you may recall from our digital audio discussion, a digital square wave is really a combination of the clock frequency with multiples, or harmonics, of that frequency. The greater the number of harmonics, the more perfect the square shape of each data pulse will be. The more square each data pulse, the more accurate the clock will be, because the transition of zero to one in each pulse establishes the clock frequency. The more accurate the clock, the less jitter, timing errors, and bit errors there will be.

Because there is no theoretical limit to the number of harmonics of a signal, we must set a practical limit. The frequency of the harmonic chosen will then be used to determine effective performance, such as cable distance. For example, choosing the ninth harmonic would allow only very short cable runs.

The third harmonic was chosen for that reason. The third harmonic is high enough to get a reasonably good digital signal but not so high that it severely affects equipment performance or cable distance. It should be noted, that the third harmonic is the *minimum* suggested frequency. Table 6-9 shows the digital frequencies, with the third harmonic frequency.

TABLE 6-9   The Third Harmonic and Wavelength

Data Rate	Clock Frequency	Third Harmonic	Third Harmonic Wavelength	Third Harmonic Quarter Wavelength
143 Mbps	71.5 MHz	215 MHz	5 ft	1 ft
177 Mbps	88.5 MHz	266 MHz	4 ft	1 ft
270 Mbps	135 MHz	405 MHz	2 ft	7 in
360 Mbps	180 MHz	540 MHz	2 ft	6 in
540 Mbps	270 MHz	810 MHz	1 ft	4 in
1.485 Gbps	750 MHz	2.25 GHz	5 in	1 in

Table 6-9 shows that a connector, cable, or other equipment intended for use at 270 Mbps (135 MHz) should be tested to 405 MHz (3 × 135). But such a part might not be acceptable for any of the higher data rates, and it is unlikely that it will be acceptable for HD.

Now we have the problem of having 50-Ω connectors, followed by 75 Ω connectors, followed by "'true'' 75 Ω connectors. Does that mean the earlier BNCs are "fake" connectors? No, it's just that they are intended for lower-bandwidth applications.

It now becomes your job to select the right BNC for your application. The key is in the questions to ask the manufacturer or distributor. If you cannot obtain the correct information or test data, then a cautious approach would be in order. As you will see, HD installations cannot be merely good enough.

Testing connectors

Changing the design of a BNC from 50 to 75 Ω is no easy task. If you look straight at the front of a BNC plug (male) and see lots of white plastic around the center pin, it is a 50-Ω BNC. If you see little or no white plastic around the center pin, if it is mostly air, then it is a 75Ω BNC. But is it 405 MHz or 2.25 GHz or what? You cannot tell simply by looking. You can look at the pin inside. If it changes its shape, if it is not consistent in its diameter, it is a lower-frequency design. If it is smooth and looks very much like a piece of center conductor in a cable, then it most likely is an HD connector.

THE ORIGINAL 75Ω BNC

It's pretty rare these days, but you might run across the original 75Ω BNC. This connector was modified for 75 Ω by changing the size of the center pin. The pin is much smaller than a regular pin. Therefore, such a connector would only be compatible with the reverse, a 75Ω jack with a small female pin. Plugging a male of these old small-pin versions into a modern BNC jack would be intermittent at best, and possibly make no contact at all. And worse would be inserting a modern 50Ω or 75Ω BNC into one of those jacks. The center pin of a regular BNC is so large, it would force the pin to spread apart or possibly break it off entirely. The way to avoid ever getting these non-compatible connectors is to be sure you ask for a "compatible BNC." All "compatible" BNCs will fit into common 50Ω or 75Ω jacks.

I emphasize "most likely" because you really can't tell just by looking. You must ask for, and receive, written data on the impedance of the connector. Believe me, no connector manufacturer today will be surprised in the least that you are asking for impedance data. Many connector manufacturers put this information in their magazine advertisements or hand out data plots at tradeshows.

You want to know that the impedance of the connector reaches at least the third harmonic of the clock. This can be displayed as a simple graph or a *Smith chart* (Figure 6-6)

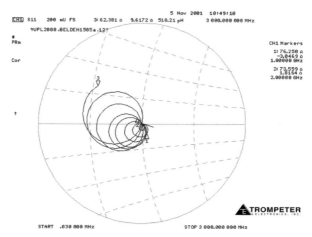

Figure 6-6 Smith chart.
(Courtesy of Trompeter Electronics.)

that shows impedance variations in a circular chart. The Smith chart is especially good at showing connectors with impedance that begins to "wander" as it gets to the high-frequency ranges. Figure 6-6 presents an example of a Smith chart showing the impedance of a BNC attached to a precision digital video cable.

Most current installations routinely use connectors measured to 2.25 GHz, even for analog applications. The reason is obvious: one connector, one strip tool, one crimp tool. No muss, no fuss. Of course, you have the economy of scale. These HD connectors are a lot more expensive, but if you buy them by the thousands, or tens of thousand for a big job, you can reduce the price considerably. It may not be at the old 50-Ω BNC price, but it will certainly be cheaper than buying two or three different connectors in small quantities. A connector tested and verified for HD applications will easily run lower-frequency digital or analog applications.

There are cables, connectors, and other passive components tested beyond 2.25 GHz. That is a good thing and indicates not only the quality but also the manufacturer's confidence in its products. Specifications that exceed the required measurement also provide the user with "headroom" by ensuring that the critical performance below this frequency is easily accomplished. Cables are available, tested, and verified to 3 GHz. Patch jacks are available, tested, and verified to 4 GHz. BNCs are available, tested, and verified to 4 GHz.

There is one additional caution. It's one thing to say, "My stuff is tested to 3 GHz"; it's another to see the data generated. Testing means nothing unless there is data to compare. Just because something is tested to some high frequency doesn't mean it's any good. It's the data that tells you that. Ask for the data.

How Many Connectors in a Wavelength?

Up to this point, connectors have had little effect on the signal. That is simply because they are not long enough to make a difference. At this point, we are getting into some seriously high frequencies.

In Table 6-9, you can clearly see we are down to an uncompressed HD third harmonic quarter wavelength of 1 in. So, how many connectors are in a quarter wavelength? Maybe one!

For those who believe the critical distance is an eighth or tenth of a wavelength, the resulting lengths are vanishingly small. It simply means that the impedance of everything is important.

The Impedance of Everything

This is where we must leave the world of separate components. The idea of considering cable, connectors, and equipment as separate pieces must be abandoned. All of these pieces must now work in unison, as a single digital entity.

We can no longer say that this part is critical and the other one is not. Everything is critical. Table 6-10 is an attempt to list, in some meaningful way, the signal path for a single digital video data stream.

Getting those bits, undistorted, to the destination is the single and ultimate aim of your system. Anything that stands in the path of their perfect delivery is a potential problem.

This table leaves out routers, distribution amplifiers, and video processing gear, all of which could be in the chain. And, yet, even with this simple example, you can see the digital signal going through twelve connectors. The total connector distance is way past the 4 in of quarter wavelength at uncompressed HD.

TABLE 6-10 The Impedance of Everything

Place	Part	Potential Problem	What To Do If It's Wrong?
Inside equipment	Integrated circuit	Ideal digital signal sent	Use a different device, swap chips
IC leads	Integrated circuit	Spacing = wrong impedance	Use a different device
Board leads	IC mounting board	Leads vary in spacing	Use a different device
Output	BNC female	50 Ω/not fully 75 Ω	Change to a 75-Ω version verified to 2.25 GHz
Cable connector	BNC male	50 W/not fully 75 Ω	Change to a 75-Ω version verified to 2.25 GHz
Interconnection	Cable	Not tested, no RL specs, varies in impedance	Change to cable tested and guaranteed as required
Cable connector	BNC male	50 Ω/not fully 75 Ω	Change to a 75-Ω version verified to 2.25 GHz
Patch panel	BNC female	Not fully 75 Ω	Change to a 75-Ω version verified to 2.25 GHz
Patch panel	Jack	Not fully 75 Ω	Change to a 75-Ω version verified to 2.25 GHz
Patch panel	Plug	Not fully 75 Ω	Change to a 75-Ω version verified to 2.25 GHz
Interconnection	Cable	Not tested, no RL specs, varies in impedance	Change to cable tested and guaranteed as required
Patch panel	Plug	Not fully 75 Ω	Change to a 75-Ω version verified to 2.25 GHz

Patch panel	Jack	Not fully 75 Ω	Change to a 75-Ω version verified to 2.25 GHz
Patch panel	BNC female	Not fully 75 Ω	Change to a 75-Ω version verified to 2.25 GHz
Cable connector	BNC male	50 Ω/not fully 75 Ω	Change to a 75-Ω version verified to 2.25 GHz
Interconnection	Cable	Not tested, no RL specs, varies in impedance	Change to cable tested and guaranteed as required
Cable connector	BNC male	50 Ω/not fully 75 Ω	Change to a 75-Ω version verified to 2.25 GHz
Input	BNC female	50 Ω/not fully 75 Ω	Change to a 75-Ω version verified to 2.25 GHz
Board leads	IC mounting board	Leads vary in spacing.	Use a different device
IC leads	Integrated circuit	Spacing = wrong impedance	Use a different device
Inside equipment	Integrated circuit	Poor digital signal received	Use a different device, swap chips

Abrupt Changes in Impedance

A microwave specialist will tell you that it's more than a question of how far is a quarter wavelength. It's also a question of abrupt changes in impedance. This is why 50-Ω BNCs can kill an HDS signal. After all, the return loss of one 50-Ω connector in a 75Ω line is about −14 dB, not horrible, and yet, there are many real-world examples where it's not just low signal. No signal gets through at all because of the abrupt change in impedance.

HORROR STORY #1,000,002

A major installer was putting in a new HD facility. He chose a high-quality connector line and everything was going fine. Because there was a very large patch-bay, the installer put in a panel with female–female feedthrough BNCs. This allowed a transition to smaller cable to make the patch panel cabling easier. Problem was, he got no signal out of the feedthrough BNCs. At first, he thought the connectors had been manufactured without pins or some other flaw. After lots of head scratching, it was obvious. The feedthroughs were 50 Ω, not 75 Ω, and the return loss let nothing through. The real problem was finding 75Ω feedthrough BNCs good to at least 2.25 GHz.

Figure 6-7 shows this effect very clearly. This test was done with a 100-ft piece of precision digital RG-59. It was cut into five pieces, and BNCs were added at the ends of each piece. The BNC connectors were probably just fine, 75 Ω to 2.25 GHz. The cable was then reassembled with BNC barrels. The four barrels were all 50Ω impedance. The result is obvious.

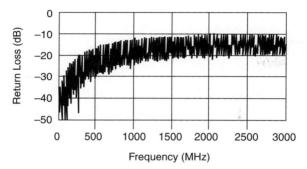

Figure 6-7 Abrupt changes in impedance.

Not only did this circuit not pass common cable expectations, −20dB return loss or better, it didn't even pass the SMPTE minimum of −15 dB return loss. We'll talk more about impedance-specific connectors in Chapter 8.

Patch Panels

We've talked about patch panels in the last few pages. It is time that we get officially to the subject. Patch panels are a critical part of any digital installation. When intended for HD applications, the jacks and plugs must stay 75 Ω to at least the third harmonic of 750 MHz, or 2.25 GHz.

There are video jacks on the market with amazing performance to 4 GHz. The funny thing is, you ask the manufacturer how it designed such a wonderful jack and they will tell you the truth: luck.

Well, to be fair, it took years of engineering experience, hundreds of millions of dollars of design and production, with a bit of luck thrown in. Why is this jack so good? Any connector manufacturer would love to know the answer. They would redesign their entire line to put in that magic dust. No doubt they eventually will figure out what makes them so good, and we will have a whole new generation of improved patch panels.

It is not unlike the manufacture of precision video coax, where all the gears, wheels, and bearings in machines are replaced on a regular basis. Which one was bad? Maybe none was bad. But the cable that then comes out of the machine is wonderful. No doubt it was this kind of approach that led to the super-jack for patch panels. Good engineering creates good products.

Digital Video on Twisted Pairs

As we have with all other signal types, we now look at running digital video on twisted pairs. Table 6-11 shows Category 5, 5e, and 6 attributes compared to the requirements for digital video. You will notice one problem right away. Category 5 and 5e cables are specified only to 100 MHz. Even 270 Mbps has a bandwidth of 135 MHz. What will these Category 5 or 5e cables do at 135 MHz? Nobody knows—they're not in the spec. But Category 6 is specified to 250 MHz, and many manufacturers will tell you specifications beyond that, to 350 MHz or even farther.

You will note that the Category 6 specification is not just crosstalk but something called PSNEXT, *power-sum near-end crosstalk*. In this specification, the three other pairs are energized with a signal and the resulting crosstalk is read on the last pairs. Each pair is so measured and the results are averaged. This test is obviously harder than just measuring pair-to-pair crosstalk. Also, there is no data point at 135 MHz, the frequency we're interested in. The closest is 155 MHz, so that is the data shown in the table. Obviously, if we get −39.5 dB of PSNEXT at 155 MHz, the number at 135 MHz will be even better, and −39.5 dB is just fine as it is.

TABLE 6-11   Digital Video on Twisted Pairs

	Digital Video	Category 5	Category 5e	Category 6
Format	Unbalanced	Balanced	Balanced	Balanced
Capacitance (pF/ft)	20	15	15	15
Impedance (Ω)	75	100	100	100
Gage	Varies	24 AWG	24 AWG	24 AWG
Shield	Yes	No	No	No
Crosstalk at 100 m (328 ft)	-30 dB at 135 MHz	No specs	No specs	−39.5 dB PSNEXT at 155 MHz

Eye Pattern

One way to tell the quality of a digital signal is by using an "eye pattern," which is generated by a test signal called a *pseudorandom bit sequence*. The resulting image on an oscilloscope, looks like an eye. When the eye is open, with no traces through the middle of the pattern, bit errors are low. When more traces fill up the eye, there are more bit errors.

How Much Crosstalk?

And how much crosstalk do we really need? Table 6-11 lists −30 dB. But ask a number of chip manufacturers that design and make ICs that decode digital and HD video and they will tell you this number is way too conservative.

What about 360 or 540 Mbps or HD 1.485 Gbps? Well, the 360 Mbps/180 MHz could be considered, because it falls in the 250 MHz of Category 6, but the other data rates are beyond the specs for even Category 6. If we had a perfect Category 6 cable, with unlimited bandwidth, the effective distance of such a cable for 1.485 Gbps/750 MHz uncompressed HD video, based on the SMTE formula, would be a whopping 90 ft.

EYES CLOSED

I once visited a prominent chip manufacturer. There I saw a perfect HD image, with no bit errors, extracted from a completely closed eye. So how much crosstalk rejection do we really need? Maybe 3 dB!

These twisted-pair cables are limited by a number of factors. First, they are 24 AWG and limited by the resistance of 24-AWG conductors. Second, the best bonded-pair Category 6 has an impedance tolerance of typically $\pm 7 \, \Omega$. That's ten times worse than precision video coax, typically $\pm 0.75 \, \Omega$. With coax, there is no such limitation. You want to go farther? Get a larger coax, with bigger conductors. At 270 Mbps/135 MHz, the "safe" distance for twisted pairs, halfway to the digital cliff, is around 225 ft. Compare that even to the smallest digital coax (750 ft), and you can see that they're not the same animal.

The limitations of twisted pairs will become even more apparent in the next chapter on broadband/CATV cables. The point is not that UTP is superior, or even equal, to coax performance, but that it is very versatile. It can run analog or digital audio. It can run analog or digital video. We'll discuss just when you might want to use category cable in Chapter 9.

VGA

VGA originally was an IBM standard. Although the cable looks identical to standard analog RGB, it is different. For one thing, these cables are for progressive-scan digital monitors, projectors, or support equipment. The sweep frequencies in these applications are much higher than analog RGB, often up to 300 or 400 MHz.

Therefore, it is important to know the attenuation of these cables at the particular sweep frequency of your equipment. Much RGB cable often is not specified in attenuation past 10 MHz, or perhaps 100 MHz. This is an indication that such cables are intended for analog RGB and, although they may work for VGA, their effective distance is unknown.

Luckily, there are a number of new multiple coax cables specified to much higher frequencies, some to 3 GHz. It is then a simple matter of comparing the attenuation at the sweep frequency of your device to its minimum input level requirements.

One thing may be immediately apparent. In large installations, such as projectors in auditoriums, standard VGA cables have too much attenuation at these high frequencies to work at the distances required. Luckily, there are bundled RG-59 and RG-6 precision video cables that offer much less loss, especially at high frequencies. If those cables are still not acceptable, you can use individual precision RG-11, but you will have to time them by hand. We will discuss timing by hand in Chapter 9.

S-VGA

The S-VGA cable includes coaxes, like VGA, with separate twisted pairs for sync signals and other control applications. The reason this sounds vague is that there is no standard for this cable. Some designers say an S-VGA cable should contain five coaxes and three twisted pair. Others say the cable should contain three coaxes and five twisted pairs. Still others say eight coaxes, or more.

If you take apart some S-VGA assemblies, you will see a great variation in design. This is one reason it is hard to buy "raw" S-VGA cable. Of course, if you buy cable through the manufacturer of the monitor or other device, it must have the appropriate design to support its applications.

7

Broadband

What Is Broadband?

Broadband is a system that was developed to deliver a very large band of frequencies to an end user. It started life, and is still commonly known as *community antenna television* or CATV. Many think this stands for "cable TV," but that is incorrect. The first CATV systems were very small, often carrying only one or two stations. Television receivers could handle only the dozen channels in the very high frequency (VHF) band, channels 2 to 13.

Electronics, who had designed and built the first television "amplifier booster" in 1947.

With the addition of the ultra-high frequency (UHF) band and constant improvements in televisions themselves, the number of channels rapidly increased. Cable systems began to add channels of their own. These started as "local access" channels, but soon expanded to include specialized channels for information and education. Satellite delivery made specialized channels profitable.

Which Channels Are Which?

There are two types of delivered channels. Off-air channels and cable/satellite channels. Tables 7-1 and 7-2 below compare these channels. The channels listed in Table 7-1 are the same for cable and off-air. Because there is a large jump between channels 6 and 7, these groups are sometimes called "low band" for channels 2 to 6 and "high band" for channels 7 to 13. Between channels 6 and 7 are a number of other broadcast services starting with FM radio, 88 to 108 MHz, air navigation, commercial aircraft communications, ham radio, land mobile, government fixed mobile, and even some space-to-earth research frequencies.

WHAT HAPPENED TO CHANNEL 1?

The FCC originally envisioned television as a series of 13 channels starting at 44 MHz. By the end of World War II, they changed their mind about Channel 1, 44–50 MHz, which they thought would interfere with other services. So it was given over to 'land-mobile' uses. No more Channel 1. Of course, these days, broadcasters are losing channels at the other end of the spectrum a lot faster!

TABLE 7-1 VHF Channels

VHF Channel	Off-Air Frequency (MHz)	Cable Frequency (MHz)
2	54–60	54–60
3	60–66	60–66
4	66–72	66–72
5	76–82	76-82
6	82–88	82–88
7	174–180	174–180
8	180–186	180–186
9	186–192	186–192
10	192–198	192–198
11	198–204	198–204
12	204–210	204–210
13	210–216	210–216

Naturally, a cable made to carry VHF television channels could be used effectively to carry any of these signals. So broadband/CATV cable is an ideal choice for FM radio reception, whether direct from a separate antenna or delivered as part of a cable television signal.

Table 7-1, with the VHF frequencies, and Table 7-2, with the UHF frequencies, show some interesting things. First, each video channel is 6-MHz wide. Within that channel the actual video signal is 4.2-MHz wide, which you might recall from Chapter 4. Second, the channels start at 54 MHz (channel 2). At this frequency, skin effect is already a serious consideration.

TABLE 7-2　UHF Channels

Off-Air UHF TV Channel	Off-Air Frequency (MHz)	Cable TV Channel	Cable Frequency (MHz)
14	120–126	A-14	470–476
15	126–132	B-15	476–482
16	132–138	C-16	482–488
17	138–144	D-17	488–494
18	144–150	E-18	494–500
19	150–156	F-19	500–506
20	156–162	G-20	506–512
21	162–168	H-21	512–518
22	168–174	I-22	518–524
23	216–222	J-23	524–530
24	222–228	K-24	530–536
25	228–234	L-25	536–542
26	234–240	M-26	542–548
27	240–246	N-27	548–554
28	246–252	O-28	554–560
29	252–258	P-29	560–566
30	258–264	Q-30	566–572
31	264–270	R-31	572–578
32	270–276	S-32	578–584
33	276–282	T-33	584–590
34	282–288	U-34	590–596
35	288–294	V-35	596–602
36	294–300	W-36	602–608
37	300–306	AA-37	608–614
38	306–312	BB-38	614–620
39	312–318	CC-39	620–626
40	318–324	DD-40	626–632
41	324–330	EE-41	632–638
42	330–336	FF-42	638–644
43	336–342	GG-43	644–650
44	342–348	HH-44	650–656
45	348–354	II-45	656–662
46	354–360	JJ-46	662–668
47	360–366	KK-47	668–674
48	366–372	LL-48	680–686
49	372–378	MM-49	686–692

continued on next page

TABLE 7-2   UHF Channels (continued)

Off-Air UHF TV Channel	Off-Air Frequency (MHz)	Cable TV Channel	Cable Frequency (MHz)
50	378–384	NN-50	692–698
51	384–390	OO-51	698–704
52	390–396	PP-52	704–710
53	396–402	QQ-53	710–716
54	402–408	RR-54	716–722
55	408–414	SS-55	722–728
56	414–420	TT-56	728–734
57	420–426	UU-57	740–746
58	426–432	VV-58	752–758
59	432–438	WW-59	758–764

Therefore, the cable used to carry the entire bandwidth of CATV/broadband has always been copper-clad steel. Such a cable is much stronger than an all-copper cable, which is especially important when cables are aerial drops. The pull strength of a standard RG-6 CATV copper-clad steel cable is 162 lb compared to an all-copper center construction pull strength of 69 lb. As we will see, there are special aerial constructions that contain extra steel wire to hold the entire cable weight, a "messenger," something that video cables do not offer.

Copper-clad steel is also important because, in most installations, the center conductor is also the connection point in the connector. Bare copper is soft and can be bent easily. Unless you are careful, this can result in an intermittent connection.

So what happened to all those other channels? Channels 70 to 83 were auctioned off and became, among other things, cell phone frequencies. Channels 60 to 69 are headed for the auction block, and will become other wireless services, soon to be followed by channels 52 to 59. The cable channels, however, are unaffected by these reassignments. Their channels continue as shown in Table 7-3.

TABLE 7-3   CATV/Broadband Channels

Channel	Frequency (MHz)	Channel	Frequency (MHz)	Channel	Frequency (MHz)
60	438–444	92	630–636	129	822–828
61	444–450	93	636–642	130	828–834
62	450–456	94	642–648	131	834–840
63	456–462	100	648–654	132	840–846
64	462–468	101	654–660	133	846–852
65	468–474	102	660–666	134	852–858
66	474–480	103	666–672	135	858–864
67	480–486	104	672–678	136	864–870
68	486–492	105	678–684	137	870–876
69	492–498	106	684–690	138	876–882
70	498–504	107	690–696	139	882–888
71	504–510	108	696–702	140	888–894
72	510–516	109	702–708	141	894–900
73	516–522	110	708–714	142	900–906
74	522–528	111	714–720	143	906–912
75	528–534	112	720–726	144	912–918
76	534–540	113	726–732	145	918–924
77	540–546	114	732–738	146	924–930
78	546–552	115	738–744	147	930–936
79	552–558	116	744–750	148	936–942
80	558–564	117	750–756	149	942–948
81	564–570	118	756–762	150	948–954
82	570–576	119	762–768	151	954–960
83	576–582	120	768–774	152	960–966
84	582–588	121	774–780	153	966–972
85	588–594	122	780–786	154	972–978
86	594–600	123	786–792	155	978–984
87	600–606	124	792–798	156	984–990
88	606–612	125	798–804	157	990–996
89	612–618	126	804–810	158	996–1002
90	618–624	127	810–816		
91	624–630	128	816–822		

Table 7-3 is not as simple and clear-cut as it seems. For instance, there are no channels 95 through 99. These are occupied by the FM band, and channels used occasionally for pay-TV services, sometimes called A-1 through A-5. There are many channels that are restricted or not advisable for use. Table 7-4 shows a few of those channels and why there could be a problem if those frequencies leaked from a CATV/broadband cable.

TABLE 7-4   Restricted Channels

Channels	Interferes With
14–16	Aeronautical communications—12.5-kHz channel offset required
18	Amateur satellite transmissions
19–20	Land mobile, emergency ship transmissions
21	NOAA distress frequencies
25–41	Aeronautical communications—12.5-kHz channel offset required
27	Emergency aircraft survival transmissions
42	Aeronautical communications—25-kHz channel offset required
43–53	Aeronautical communications—12. 5-kHz channel offset required
88–89	Interferes with set-top box intermediate frequencies (IFs)
98–99	Aeronautical communications—25-kHz channel offset required
145	Second local oscillator in some televisions
151–153	First intermediate frequency in some televisions

You can see why the FCC is so insistent about the leakage of signals from cable. There are specific laws and requirements for emission that cables and equipment must pass if it is intended for CATV/broadband applications.

Channels 42 through 116 are often reserved for delivery of digital channels. There is demand for even more channels, up to 500. The spectrum for these channels has not been agreed upon or allocated. The bandwidth of 3 GHz for the entire spectrum will put added pressure on cable manufacturers to produce lower-loss cable versions.

You Heard It Here First

There is one technology that could give fiber a run for the money—room-temperature superconductors. Imagine a wire with no resistance. If you get regular wire cold enough, close to absolute zero, −273°C (−459°F), then it too has no resistance. Recent work has versions using carbon "bucky balls" up to a summery −249°F. Eventually, they may get to room temperature. This would mean wire, connected much the same as current copper wires, but with no resistance. You could send a signal across the room, or across the world, with equal level. No conversion from electrical to optical would be necessary. Because there is no loss, there would be no attenuation at any frequency; there would be unlimited bandwidth. Currently, they have produced tiny fibers smaller than hair and very short. They are so thin, in fact, that any signal put on such a wire creates a huge magnetic field that blows up the wire on the quantum level. So, if you hear of room-temperature superconductors coming on the market, see your stockbroker!

Lower-loss cables essentially would lead to ultra-high velocity cable designs at much greater cost and could usher in fiber to the home, based solely on the bandwidth requirements of new services. Because the majority of broadband distribution is on single-mode fiber and then converted to copper cables for the "drop" to each home, fiber is often much closer to the customer than one might realize. Splitting and delivering signals on fiber, however, which is being done in a few experimental communities, presents major challenges compared to copper cables. Single-mode fiber, with bandwidth limited only by the equipment attached, makes ideal sense for long-run high-bandwidth delivery. Because the signal is optical, it is impervious to electromagnetic noise of any kind.

WHAT WERE THEY THINKING?

When the cable companies put together their channel order, they must have done it with a pair of dice. A-14 through I-22 are midband channels. J-23 through W-36 are superband channels. AA-37 through ZZ-62 are hyperband channels. And 95-99 jump all the way back to the FM channels. Makes your head spin!

In the copper realm, a signal can be split and distributed passively, with a constant-impedance splitter, or actively, with a distribution amplifier. This is a very expensive option with fiber if the signal is kept in the "optical" mode. Most often, active splitting requires that the signal on the fiber first be converted from optical to electrical, split, and then re-amplified, and each leg converted back to optical. Even though this is considerably cheaper than keeping the signal optical, the expense of either proposition is considerable. Therefore, it is likely that the progression of fiber to the home will gain momentum only when copper cables can no longer deliver appropriate levels at appropriate bandwidths.

Cable also uses other channels for interactive services. They are listed in Table 7-5.

TABLE 7-5   A and T Channels

Cable Channel	Frequency (MHz)
A-3	102–108
A-4	96–102
A-5	88–96
A-8	72–76
T-7	6–12
T-8	12–18
T-9	18–24
T-10	24–30
T-11	30–36
T-12	36–42
T-13	42–48

A and T channels were intended for interactive services, that is, signals and data from the end user to the cable company, the so-called "upstream" signals. The problem is that some of the frequencies that were chosen, especially the T channels, are very low, below 54 MHz (channel 2).

Cable amplifiers, with a typical gain of 25 to 30 dB, are inserted to overcome the losses at high frequencies. In-line filter/attenuators are added so that the T channels do not overpower the amplifiers in the reverse direction. Imagine thousands or tens of thousands of customers sending their own particular data streams to the head end. These signals are repeatedly combined and run through dozens of amplifiers. The result is uncontrollable noise and very limited bandwidth from each end user. Fiber optic return feeds have much less equipment in the line, are immune from RF noise, and may be the final answer for effective interactive services.

You may recognize the A-channels in Table 7-5 as being the same frequencies as the FM band. Using these channels for video services can create interference between the cable and existing FM signals. If the cable leaks or emits its own signal at that frequency, it can affect FM reception on other radios in the home or nearby televisions. Existing FM stations, if nearby with high signal strength, can interfere with signals on those channels.

Hence, A and T channels are not in great use. There were great plans in the 1970s and 1980s to have all sorts of interactive services. These included the ability of the user to change to a different camera view during a sports game, play interactive games with other viewers, on-screen wagering, contests, and many other two-way applications. The limitations of the lower channels and the structure of the existing cable television and FM radio systems made those cost prohibitive. Plus, none of these offered services was the "killer app" that every cable viewer would die to have.

Some of those applications did survive in smaller, simpler versions. The downstream connection can be made on the cable, whereas the upstream (return) connection is often made on a telephone line with a dial-up mechanism in the cable decoder box.

How Broad Is Your Band?

Some cable and satellite providers have begun to offer other services in addition to television programming, such as Internet access. Of course, they have a huge pipe, huge in the electronic bandwidth sense, and can easily offer very fast download services, often 256 Kbps or more.

The television offering is presented identically to every customer, whereas each Internet customer wants to download something different. During peak hours, huge amounts of bandwidth can be used up. The total bandwidth on the system is based on the number of customers and their individual needs. As far as the coax itself is concerned, the only thing we're interested in is the ultimate bandwidth required to deliver.

The Cable Box

As shown in Tables 7-1 to 7-3 the cable channels and the off-air channels do not always coincide. Special channels may be encoded to further protect them from being stolen by non-paying customers. Hence, a "cable-ready" television can tune and display only the standard broadcast-style signals. A cable-ready television cannot decode the special channels with movies, sports, and pay-per-view fare. Of course, a cable-ready television cannot decode any non-television service, such as Internet access.

Some cable channels are off-set by the cable company as shown in Table 7-4. This involves shifting the frequency of the channel to minimize intermodulation (IM) between the cable signal and the direct off-air signal. This IM distortion shows up as visible "beats" in the television picture. The on-air broadcast signals, especially in the VHF spectrum, correspond exactly with the cable channels, as can be seen in Table 7-2. Unless the cable, connectors, and other hardware offer excellent shielding and noise rejection, ghost images can be easily generated on a television from the cable and on-air signals being received.

The ghosting, in that instance, is caused by the fact that the two paths, through the air and the cable, are not the same length. This means the two identical images arrive at different times and are both displayed. Improvements in connectors and cable shield effectiveness have solved most of these problems.

Which Cable to Use?

The coax cabling required for broadband is the same as for any other multichannel television application. The cables differ based on where and how they are installed and what the required bandwidth will be.

SCTE

The Society of Cable Telecommunications Engineers originally was the Society of Cable Television Engineers. When cable companies began to offer other services besides television, the society changed the name to reflect the difference.

The Society of Cable Telecommunications Engineers
140 Philips Road, Exton, PA 19341.
Phone: 610.363.6888/800.542.5040,
Fax: 610.363.5898
www.scte.org

There are a number of variations on a theme for CATV/broadband coax. The voluntary industry specifications under which they are designed and constructed have been developed by the SCTE.

Could you make a better CATV/broadband cable? With high velocity gas-injected foam, you could. However, such a cable would have to have different dimensions to stay at 75 Ω. Such a super-cable would not be compatible with existing connectors. Therefore, most cables made for CATV/broadband applications are very close in design, so that they can fit standard connectors.

This does not mean that all cables are the same. The crush testing shown in Tables 6-3 and 6-4 is based on testing done originally on cable television constructions. There can be dramatic differences in quality between different manufacturers, based on this one factor. These cables may look the same when compared to the SCTE spec, but they are not the same in quality.

How Broadband/CATV Cables Are Sold

If you are a designer or installer of broadband/CATV systems, you know what the problem is. There are a few very big players in the broadband/CATV world. They buy cable by the millions or tens of millions of feet. Pricing for them is incredibly low, often with only a tiny margin per foot to the manufacturer. Of course, the manufacturer hopes to make up for this difference in volume.

What this means is that CATV/broadband coax is made as close to the minimum spec as possible. A one-cent difference in the overall cost of a cable can translate into hundreds of thousands of dollars for a very large order. So this type of coax is really the opposite of HD broadcast coax. Broadcast coax is made with consistency and performance as the #1 and #2 goals. Price, in that case, is #3. With broadband/CATV coax, it's just the reverse: price is #1, consistency is #2, and quality is #3.

This order of importance is more apparent when we look at the various types of broadband/CATV cables available. The following tables list virtually every possible type. Nevertheless, every cable has some factors in common. For instance, these are 75Ω coaxes. All use copper-clad steel center conductors, and many now use high-velocity nitrogen-gas–injected foam dielectrics.

The first difference you will note is that the "RG" designator used so far has been replaced by "Type." This is because of the addition of Type 7 coax, which has no relationship to the original RG system. Further, these modern designs are so far from the original RG designs that the relationship between the original military cables and these cables is, at best, tenuous.

These cables follow a simple progression. There are different braid coverages, from 40 to 80 percent. Sometimes you can find 90 percent or higher braid coverage, but this is at a higher cost and not often used.

Shield Coverage

Foil shields are highly effective at all channel frequencies. The braid shield has three purposes. First, it is the path to ground for the noise picked up by the foil. Second, it gives the connector at each end something to hold onto. Third, it adds a small amount of strength to the tension the cable can withstand.

Very little shield effectiveness is provided, or required, from these braid shields. It is more a question of reliability, at least from the connector perspective. This is why braid coverage higher than 90 percent adds performance but is a significant addition in cost. Nowhere is the price–performance ratio more keenly felt than in CATV/broadband constructions. Any change suggested in these cables must be justified by a major improvement in performance for any added cost. Otherwise, the change had better reduce the cost.

Cost also dictates, to a great extent, how these cables are made, and what they are made of. For instance, braid shields in CATV/broadband cables are almost always aluminum braid. Aluminum braid is inexpensive and easy to manufacture. Because the foil underneath is also aluminum, there is no corrosive potential between these two layers. Broadband/CATV cable is often placed outdoors and exposed to all kinds of weather, so the potential for corrosion is a major consideration.

Braidless Coax

Any manufacturer will tell you that braiding is the most labor-intensive step in the manufacture of CATV/broadband coaxes. If a cable could be made without a braid, it would be a major breakthrough in cost reduction, and many have tried to produce braidless coax.

But these cables failed in the marketplace for a number of reasons. If manufacturers reduced the amount of braid, then the resistance of the braid and the ability to drain off noise collected on the foil shield was seriously compromised.

If multiple large wires were used to decrease the resistance, it became very hard to put on a connector. Reliability of the connection was compromised. So here is your billion-dollar idea for the week. Come up with a way to make a coax that has the same resistance as a braid, without a braid, but make it easy and reliable to connect.

Braid Resistance

The resistance of aluminum braid is higher than that of copper braid. That simply means you can pick a higher braid coverage if you want lower resistance. It is interesting to note that cables with very high braid coverage, 90 percent or more, are rare in broadband constructions. In general, 80 percent coverage is more than adequate to meet the needs of these cables, so that is generally the maximum braid coverage available.

Shield Effectiveness

The key to shield effectiveness in CATV/broadband cables is a measurement called *transfer impedance*. A signal is generated on one side of a shield and the result is read on the interior of the shield. The voltage drop generated on the other side of the shield and the ratio between them, as shown in Figure 7-1, comprise the transfer impedance.

$$Z_t = \frac{1}{I_o} \times \frac{dV}{dX}$$

Figure 7-1   Formula: transfer impedance.

This test method was an updated version of one used in Europe and was originally proposed by Kenneth Simmons in 1973. Z_t is the transfer impedance in ohms. I_o is the current of the signal on one surface. The ratio dV/dX is the voltage produced by that current because of the resistance of the shield. The resistance of the shield is based on the length, X, of the test cable.

The lower the value of transfer impedance, the more effective the shield. Of course, this is a laboratory test. So far, there are no field testers sophisticated enough to run measurements like this in the real world. What transfer impedance does show, at least theoretically, is the difference between the external real world and the internal world of the dielectric and center conductor.

Table 7-6 lists different shield types.

TABLE 7-6 CATV/Broadband Shields

Construction	Details
Bonded	Foil glued to the dielectric
Shorted foil	Foil with one edge turned so that it completes the shield by connecting to the other edge when the foil is wound around the core.
Tri-shield	Foil-braid-foil
Quad shield	Foil-braid-foil-braid

Table 7-7 shows typical transfer impedance values in milliohms per meter based on the different shield types.

Quad-shielded cable is often specified as the ultimate in shield performance; it is not. High-coverage braid, even with a single foil, will outperform a standard quad-shield. Of course, if you can get high-coverage braids in a quad, it would do better still. The standard quad configuration, two braids of 60 and 40 percent coverage, are shown in Table 7-7. These two braids in a quad-shield do not add up to 100 percent braid coverage, as some incorrectly believe. Each has its own transfer impedance and the resultant leakage is more than a reasonably high single braid.

TABLE 7-7 Transfer Impedance

Construction Type 6	5 MHz	10 MHz	50 MHz	100 MHz	500 MHz
Bonded + 60% braid	20	15	11	20	50
Tri-shield + 60% braid	3	2	0.8	2	12
Quad-shield + 60%/40% braid	2	0.8	0.2	0.3	10
Tri-shield + 80% braid	1	0.6	0.1	0.2	2
Bonded + 95% braid	1	0.5	0.08	0.09	1

Many major cable users believe that very high-coverage braids, such as the 95 percent braid shown, are simply not worth the money. Although they are demonstrably superior, these major cable users feel that the cost of manufacturing these high-coverage braids simply cannot be justified by such a slight improvement in performance.

Cable companies buy cable in the millions of feet and are conscious of every penny difference in price. You, as a designer, installer, or contractor, are not held to the same limitation. In fact, you may wish to use higher braid coverages, at least 80 percent, so that you can differentiate your installation from the standard.

The Ultimate CATV/Broadband Install?

Some installers even use precision video cable for CATV/broadband installations. Features such as 95 percent copper braid coverage, guaranteed superior return loss, and sweep testing to 3 GHz, can make such a cable dramatically better than even specialized broadband cables. Just be sure that the dimensions of the precision video cable are very close, if not identical, to those of broadband/CATV dimensions. There are RG-6 (Type 6), RG-59 (Type 59), and RG-11 (Type 11) cables that would be compatible. Many of the older analog video designs are incompatible with broadband/ CATV connectors.

Of course, the cost of a precision video cable is many times higher than even specialty CATV/broadband cables. But, if you're aiming for the top-of-the-top, you can use precision video coaxes. The only disadvantage is the all-copper center that is a lot softer than copper-clad steel. So you have to be a bit more careful when a bare solid copper center conductor is the connecting part of a CATV/broadband cable.

Using precision video coax for CATV/broadband also means you can use the same cable to run baseband video, unbalanced analog audio, S/PDIF, and AES3-id. All you might need is to buy a few different colors, or do a good job at labeling one color of cable. The economy of scale means you would buy more of one kind of cable, thus lowering the price per foot, instead of small quantities of three or four different cables.

Basic Sizes

The four basic sizes of cable are shown in Table 7-8. Adding multiple shields, special layers for waterblocking or direct burial, or messenger wires for aerial installation may add to the diameter.

TABLE 7-8   Basic Sizes

Type	Center Conductor	Diameter (in)
Type 59	20 AWG	0.240
Type 6	18 AWG	0.270
Type 7	16 AWG	0.320
Type 11	14 AWG	0.400

Cable Features

Table 7-9 is a list of the basic features and what they mean. "Tested and verified" does not mean, as with precision video cable, that every reel of CATV/broadband cable is tested. This would be a considerable expense for which the usual CATV/broadband customer would not pay. More likely, one reel in every master reel is tested. A master reel in cable manufacturing is usually 25,000 ft. That would mean that one roll in every 25 is tested. In fact, many manufacturers will only test one reel in 100 to keep cost down.

TABLE 7-9   Cable Features

Feature	What It Means
1 GHz	Cables are tested and verified to 1 GHz
2.25 GHz	Cables are tested and verified to 2.25 GHz
Messenger	An extra steel wire is included to suspend the cable

Therefore, you are trusting the manufacturer for the performance of the other 99 reels, so consistency matters. If the manufacturer is consistent, it is easier to trust that the other reels are nearly identical to the test reel.

Tables 7-10 and 7-11 show the most common types of cables. Omitted are dual-coaxes, direct burial versions, dry or gel-filled waterblocking, and other special constructions. These tables do not mean a particular manufacturer will make every one of these types, but these variations are possible.

There are a few basic themes and variations on each theme (Table 7-12). As these cables get larger, to Type 11 for instance, it is less likely that all variations are available. Type 6 is the key size. Virtually all cabling in consumer installations is Type 6. And virtually all the constructions shown are common in Type 6.

You could always have a special cable made to your specifications. If you require a waterblocked type, such as Type 11 (Table 7-13), with a messenger, any one of a number of manufacturers would be more than happy to make it. The only consideration would be the minimum purchase, usually between 10,000 and 25,000 ft.

Hard Line

Hard line is large coax with a solid center conductor and an extruded or rolled aluminum shield. It is extremely stiff and very difficult to bend and install. It used to be the standard for long runs. Until the advent of single-mode fiber, hard line was the standard trunk cable for delivery of CATV signals.

Hard line comes in a number of sizes including 0.625, 0.750, 0.875, 1, and 1.25-in. Size, as with any coax, determines loss, so the larger hard line cable can carry signals farther before an in-line CATV amplifier is needed to boost the signal. In the rare instance where hard line is part of a building install, the 0.750-in size is the most common. You most likely would match whatever size of hard line is coming into the building and continue to the termination point.

TABLE 7-10 Type 59 Cable

Type	Conductor	Shield	Swept To	Addition
59	20 AWG	100% foil + 40% braid	1 GHz	
59	20 AWG	100% foil + 60% braid	1 GHz	
59	20 AWG	100% foil + 80% braid	1 GHz	Messengered
59	20 AWG	100% foil + 40% braid	1 GHz	Messengered
59	20 AWG	100% foil + 60% braid	1 GHz	Messengered
59	20 AWG	100% foil + 80% braid	1 GHz	
59	20 AWG	100% foil + 40% braid	2.25 GHz	
59	20 AWG	100% foil + 60% braid	2.25 GHz	
59	20 AWG	100% foil + 80% braid	2.25 GHz	Messengered
59	20 AWG	100% foil + 40% braid	2.25 GHz	Messengered
59	20 AWG	100% foil + 60% braid	2.25 GHz	Messengered
59	20 AWG	100% foil + 80% braid	2.25 GHz	
59	20 AWG	100% foil + 40% braid + 100% foil	1 GHz	
59	20 AWG	100% foil + 60% braid + 100% foil	1 GHz	
59	20 AWG	100% foil + 80% braid + 100% foil	1 GHz	Messengered
59	20 AWG	100% foil + 40% braid + 100% foil	1 GHz	Messengered
59	20 AWG	100% foil + 60% braid + 100% foil	1 GHz	Messengered
59	20 AWG	100% foil + 80% braid + 100% foil	2.25 GHz	
59	20 AWG	100% foil + 40% braid + 100% foil	2.25 GHz	
59	20 AWG	100% foil + 60% braid + 100% foil	2.25 GHz	
59	20 AWG	100% foil + 80% braid + 100% foil	2.25 GHz	
59	20 AWG	100% foil + 40% braid + 100% foil	2.25 GHz	Messengered
59	20 AWG	100% foil + 60% braid + 100% foil	2.25 GHz	Messengered

59	20 AWG	100% foil + 80% braid + 100% foil	2.25 GHz	Messengered
59	20 AWG	Foil + 40% braid + foil + 60% braid	1 GHz	Messengered
59	20 AWG	Foil + 40% braid + foil + 60% braid	1 GHz	
59	20 AWG	Foil + 40% braid + foil + 60% braid	2.25 GHz	Messengered
59	20 AWG	Foil + 40% braid + foil + 60% braid	2.25 GHz	

TABLE 7-11 Type 6 Cable

Type	Conductor	Shield	Swept To	Addition
6	18 AWG	100% foil + 40% braid	1 GHz	
6	18 AWG	100% foil + 60% braid	1 GHz	
6	18 AWG	100% foil + 80% braid	1 GHz	Messengered
6	18 AWG	100% foil + 40% braid	1 GHz	Messengered
6	18 AWG	100% foil + 60% braid	1 GHz	Messengered
6	18 AWG	100% foil + 80% braid	1 GHz	
6	18 AWG	100% foil + 40% braid	2.25 GHz	
6	18 AWG	100% foil + 60% braid	2.25 GHz	
6	18 AWG	100% foil + 80% braid	2.25 GHz	
6	18 AWG	100% foil + 40% braid	2.25 GHz	Messengered
6	18 AWG	100% foil + 60% braid	2.25 GHz	Messengered
6	18 AWG	100% foil + 80% braid	2.25 GHz	Messengered
6	18 AWG	100% foil + 40% braid + 100% foil	1 GHz	
6	18 AWG	100% foil + 60% braid + 100% foil	1 GHz	
6	18 AWG	100% foil + 80% braid + 100% foil	1 GHz	
6	18 AWG	100% foil + 40% braid + 100% foil	1 GHz	Messengered
6	18 AWG	100% foil + 60% braid + 100% foil	1 GHz	Messengered
6	18 AWG	100% foil + 80% braid + 100% foil	1 GHz	Messengered
6	18 AWG	100% foil + 40% braid + 100% foil	2.25 GHz	
6	18 AWG	100% foil + 60% braid + 100% foil	2.25 GHz	
6	18 AWG	100% foil + 80% braid + 100% foil	2.25 GHz	
6	18 AWG	100% foil + 40% braid + 100% foil	2.25 GHz	Messengered
6	18 AWG	100% foil + 60% braid + 100% foil	2.25 GHz	Messengered

6	18 AWG	100% foil + 80% braid + 100% foil	2.25 GHz	Messengered
6	18 AWG	Foil + 40% braid + foil + 60% braid	1 GHz	Messengered
6	18 AWG	Foil + 40% braid + foil + 60% braid	1 GHz	
6	18 AWG	Foil + 40% braid + foil + 60% braid	2.25 GHz	Messengered
6	18 AWG	Foil + 40% braid + foil + 60% braid	2.25 GHz	

TABLE 7-12 Type 7 Cable

Type	Conductor	Shield	Swept To	Addition
7	16 AWG	100% foil + 40% braid	1 GHz	
7	16 AWG	100% foil + 60% braid	1 GHz	
7	16 AWG	100% foil + 80% braid	1 GHz	
7	16 AWG	100% foil + 40% braid	1 GHz	Messengered
7	16 AWG	100% foil + 60% braid	1 GHz	Messengered
7	16 AWG	100% foil + 80% braid	1 GHz	Messengered
7	16 AWG	100% foil + 40% braid	2.25 GHz	
7	16 AWG	100% foil + 60% braid	2.25 GHz	
7	16 AWG	100% foil + 80% braid	2.25 GHz	
7	16 AWG	100% foil + 40% braid	2.25 GHz	Messengered
7	16 AWG	100% foil + 60% braid	2.25 GHz	Messengered
7	16 AWG	100% foil + 80% braid	2.25 GHz	Messengered
7	16 AWG	100% foil + 40% braid + 100% foil	1 GHz	
7	16 AWG	100% foil + 60% braid + 100% foil	1 GHz	
7	16 AWG	100% foil + 80% braid + 100% foil	1 GHz	
7	16 AWG	100% foil + 40% braid + 100% foil	1 GHz	Messengered
7	16 AWG	100% foil + 60% braid + 100% foil	1 GHz	Messengered
7	16 AWG	100% foil + 80% braid + 100% foil	1 GHz	Messengered
7	16 AWG	100% foil + 40% braid + 100% foil	2.25 GHz	
7	16 AWG	100% foil + 60% braid + 100% foil	2.25 GHz	
7	16 AWG	100% foil + 80% braid + 100% foil	2.25 GHz	
7	16 AWG	100% foil + 40% braid + 100% foil	2.25 GHz	Messengered
7	16 AWG	100% foil + 60% braid + 100% foil	2.25 GHz	Messengered

7	16 AWG	100% foil + 80% braid + 100% foil	2.25 GHz	Messengered
7	16 AWG	Foil + 40% braid + foil + 60% braid	1 GHz	Messengered
7	16 AWG	Foil + 40% braid + foil + 60% braid	1 GHz	
7	16 AWG	Foil + 40% braid + foil + 60% braid	2.25 GHz	
7	16 AWG	Foil + 40% braid + foil + 60% braid	2.25 GHz	Messengered

TABLE 7-13 Type 11 Cable

Type	Conductor	Shield	Swept To	Addition
11	14 AWG	100% foil + 40% braid	1 GHz	
11	14 AWG	100% foil + 60% braid	1 GHz	
11	14 AWG	100% foil + 80% braid	1 GHz	
11	14 AWG	100% foil + 40% braid	1 GHz	Messengered
11	14 AWG	100% foil + 60% braid	1 GHz	Messengered
11	14 AWG	100% foil + 80% braid	1 GHz	Messengered
11	14 AWG	100% foil + 40% braid	2.25 GHz	
11	14 AWG	100% foil + 60% braid	2.25 GHz	
11	14 AWG	100% foil + 80% braid	2.25 GHz	
11	14 AWG	100% foil + 40% braid	2.25 GHz	Messengered
11	14 AWG	100% foil + 60% braid	2.25 GHz	Messengered
11	14 AWG	100% foil + 80% braid	2.25 GHz	Messengered
11	14 AWG	100% foil + 40% braid + 100% foil	1 GHz	
11	14 AWG	100% foil + 60% braid + 100% foil	1 GHz	
11	14 AWG	100% foil + 40% braid + 100% foil	1 GHz	Messengered
11	14 AWG	100% foil + 60% braid + 100% foil	1 GHz	Messengered
11	14 AWG	100% foil + 80% braid + 100% foil	1 GHz	Messengered
11	14 AWG	100% foil + 40% braid + 100% foil	2.25 GHz	
11	14 AWG	100% foil + 60% braid + 100% foil	2.25 GHz	
11	14 AWG	100% foil + 80% braid + 100% foil	2.25 GHz	
11	14 AWG	100% foil + 40% braid + 100% foil	2.25 GHz	Messengered
11	14 AWG	100% foil + 60% braid + 100% foil	2.25 GHz	Messengered

11	14 AWG	100% foil + 80% braid + 100% foil	2.25 GHz	Messengered
11	14 AWG	Foil + 40% braid + foil + 60% braid	1 GHz	
11	14 AWG	Foil + 40% braid + foil + 60% braid	1 GHz	Messengered
11	14 AWG	Foil + 40% braid + foil + 60% braid	2.25 GHz	
11	14 AWG	Foil + 40% braid + foil + 60% braid	2.25 GHz	Messengered

The center conductors of these hard-line cables are made of copper-clad aluminum. The dielectric is extruded foam or fused disks of dielectric. The shield is an extruded solid aluminum tube or rolled and welded. The rolled and welded variety can be made significantly thinner than the extruded type. Waterblocking is a serious issue for hard line because it is often used outdoors. The most critical part of waterblocking is at the connectors where failures commonly occur.

Putting connectors on hard line is hard work. Just cutting and stripping these cables require specialized tools. If you are inside a building, you might consider a transition to Type 11 or even smaller coaxes. Hard line is only necessary where you have extremely long runs or very wide bandwidth and cannot afford the attenuation of smaller cables.

Sag and Span

Cables that employ a messenger, an extra steel wire outside the coax, are intended for suspended aerial installations. When so used, the requirements for tension versus length and sag are essential to know.

Table 7-14 shows generic values for three sizes of messenger wires. Some manufacturers can exceed those generic numbers by a considerable percentage. Contact the manufacturer for details on their specific products. Table 7-14 shows conservative estimates for the design of most CATV/broadband installations.

Table 7-14 also shows common values for sag and span, with the maximum tension on the messenger. Sag is the distance a cable will drop due to its own weight. The amount of sag is affected by the distance, or span, of the cable. You can read the tension with the use of a fish scale or other similar device. Increasing the sag reduces the tension, and reducing the sag increases the tension.

TABLE 7-14   Sag and Span

Messenger size	0.062 in	0.072 in	0.083 in
Sag 2 ft	47 ft	55 ft	62 ft
Sag 3 ft	57 ft	66 ft	75 ft
Sag 4 ft	66 ft	76 ft	86 ft
Sag 5 ft	73 ft	86 ft	96 ft
Sag 6 ft	80 ft	93 ft	105 ft
Sag 7 ft	87 ft	101 ft	113 ft
Sag 8 ft	93 ft	108 ft	121 ft
Maximum tension	271 lb	365 lb	460 lb

Connectors

The standard connector for CATV/broadband installations is the F connector. There are three common types. The most common is the crimp style. Table 7-15 lists the generic steps required to put on a standard crimp-type F. Of course, for a specific connector type, you should always contact the connector manufacturer to obtain cut and crimp dimensions.

TABLE 7-15   Cable Preparation

1.	Blunt cut cable to the correct length
2.	If connector has a separate crimp ring, put it on now
3.	Use a cable stripper to remove jacket
4.	Use a cable stripper to remove dielectric
5.	Push braid back
6.	Insert connector body
7.	Push braid back over connector
8.	Push crimp ring over body
9.	Make sure appropriate length of center conductor is exposed
10.	Check that dielectric is up as far as possible in the connector
11.	Crimp the crimp ring

You strip the coax cable to the dimensions set by the connector manufacturer. The length of exposed center conductor should be between $1/4$ and $1/2$ inches. For added strain relief when used indoors, there are plastic boots available for some of these connectors.

An F Is an F?

I have a friend who, just on a lark, started to put F connectors in his test jig for BNCs. Of course, these are all 75-Ω connectors, right? He was amazed. Many F connectors are a long way from 75 Ω. One measured 53 Ω! If you're doing serious CATV/broadband installations, maybe you should get impedance plots or Smith charts from the manufacturer. Of course, they're not used to this and you might throw them for a loop, but ask about it anyway.

The second style of connector is the compression-fit version, which often is considerably smaller. This two-piece connector clamps the braid and squeezes the two parts together. It is especially effective when installing hundreds of cables where a regular-size connector takes up too much room.

Drop Cable Failures

There is evidence that the F connector is a major contributor to drop cable failure. The sharp edges in the connector design contribute to RF radiation at higher frequencies. The wide range of center conductor lengths leads to loss of retention and intermittent operation due to expansion and contraction. The

> return-loss requirement on broadband/CATV cable is
> −20 dB SRL (remember SRL?). As we go into the
> 500-channel world, real return loss will be required
> on cable and connectors.

The third connector is the captured-pin F connector. This connector is much larger than the other F connectors. It contains a center pin, into which the center conductor of the cable is inserted. The center conductor makes contact by pressure only and is not crimped or otherwise connected. The sleeve of the connector is crimped or clamped into place. The captured-pin F is more rugged than other connectors, but the connection to the cable is less rugged. It looks nicer, more like a real connector, such as a BNC, and is often used where appearance is important. Its dimensions are more precise than those of standard F connector.

If, for some reason, you are using stranded video cable, you cannot use it with a generic F connector. Stranded cable has much higher impedance variations and return loss and cannot carry the bandwidth required for broadband without significant loss. However, surveillance cameras, especially those that move all the time, require stranded cable and often use the F as the connector of choice. This is a good time to use a captured pin F.

Do not try to solder a stranded center conductor to insert into a female F. It is almost impossible to get the appropriate dimensions and the resulting wire is even softer with the addition of the tin/lead solder, so the result is almost always an intermittent connection. Further, the tin-lead solder will have significant corrosive potential with the copper and steel in the female receptacle, especially if used outdoors. Use a captured-pin F or, if you're desperate, put on a BNC and an F adapter.

Distribution

Distribution of CATV/broadband signals has been done for decades in a standard "trunk and drop" style. The trunk cable carries the signal from a cable head end or CARS microwave receive location and runs through the area to be served. Drop cables deliver the signal to each user. These days, the trunk cable is single-mode fiber that is converted from optical to electronic, and then drop cables, most commonly RG-6, deliver the signal to the user. This is called "fiber-to-the-node" architecture and results in amplifier cascades of perhaps one trunk amplifier, one bridge amplifier, and two line-extender amplifiers. This minimal amplifier cascade makes cable bandwidth of 1 GHz or beyond practical.

HIGH-DEFINITION 1968

I remember walking into one of my local TV stations in 1968 and being stunned by the image on the monitor. I remember it was an episode of The Beverly Hillbillies, and it was amazing. If they could have delivered that quality to the home, you would have thought it was HD. And no wonder, because that image had 700-line resolution, whereas my TV at home barely had 400, and many televisions were soon reduced to 200 lines to remove the inherent noise in the signal. We traded detail for low noise.

Prior cable architecture, before fiber backbones, required trunk amplifier cascades as high as 30, which typically limited the highest frequency to about 450 MHz. Fiber backbones have made 750 MHz and 1 GHz or beyond practical fiber-to-the-node designs.

Distribution within a home or commercial building is a miniature of the distribution scheme from the cable provider, except that it is all-coax. It is rare that fiber is used in an installation There are only a few, mostly experimental, communities that deliver fiber to the end user. If, however, fiber is delivered to your installation site, you can extend the fiber to any part of the building before converting from optical to electronic.

You will still use coax trunk-and-drop cables in commercial buildings or home installations, no matter how the signal is delivered to the building. The signal will be amplified and then split into separate coaxes for each destination. Often, the amplification and splitting can be done with one unit, a distribution amplifier.

STRING OF IMAGES

I once sold a lot of in-line CATV amplifiers and passive splitters to a group that was doing an Academy Awards banquet. They rented a truckload of televisions from the local rental house and just wanted them hooked up. I showed up to help and realized that they intended to daisy-chain all the televisions. Hey, it's their gig. So I led the feed to the first TV, put in a two-way splitter, and put a cable to the next TV. Sometimes I used a three- or four-way splitter if there were a lot of TVs in one place. If the signal began to get noisy, I would throw in an in-line amp and keep on going. I didn't expect to get very far doing this and was suitably amazed when all the TVs were hooked up and the image on the last one was perfectly acceptable. I would guess there were 50 sets off of 20 splits with an amp every four or five splits, for 300 ft.

In a hospital or hotel, for instance, it is common to have a trunk cable travel from the source and pass every floor. Distribution amplifiers have a feedthrough port that can be used to attach the source trunk cable and then continue on to feed another distribution amplifier on the next floor. If amplification is not offered for this trunk port, each split will drop the signal by −3 dB, plus the loss of two connectors.

If the source signal is strong enough, it can feed many floors before the level has fallen to an unusable level. The key is to not overdrive the inputs to the first distribution amplifier, but have enough level after all the splits for the last amplifier. Because there is an amplifier on each split, you can adjust for the loss and feed an appropriate level to each television.

You can also use separate single-channel amplifiers and passive splitters. It is a bit more complicated and not as simple as the distribution amplifier, but it can be more versatile.

Level and Noise

The problem with the trunk-and-drop approach is noise. When you split the signal, and lose −3 dB, the signal-to-noise ratio also drops by −3 dB. When you amplify that signal by +3 dB, the noise floor is also amplified, so the noise rises by 3 dB. After a number of splits, you can have adequate level, but terrible signal-to-noise ratios.

Older cable systems were lucky to provide viewers with carrier-to-noise (C/N) ratios of better than 40 dB. Modern-day analog systems provide C/N ratios of 45 to 50 dB. FCC rules now require a minimum of 44 dB. Digital systems can deliver a C/N of 60 dB or better.

That may be the reason HD is taking so long to be integrated into cable systems. Once viewers see an HD image, things will never be the same and the average video image will be unacceptable. It will be the same as the CD in the audio world that changed listeners from 60 to 90 dB signal-to-noise ratio levels almost overnight.

Digital Systems

Things will change when signals are delivered and then distributed in the digital mode. This is already an option for many broadband companies, although the price of the set-top box to do the conversion is a lot more expensive than the old analog box. Still, the image is so much cleaner that customers are demanding it.

Interestingly, the bandwidth of a digital channel is exactly the same, 6 MHz, as an analog channel. Even an HD image fits into a 6-MHz channel. The reason is that the signal is dramatically compressed. The HD signal, 1.485 Gbps/750 MHz in the studio, is transmitted over the air or delivered by cable as a 19.4-Mbps signal, a compression ratio of more than 76 to 1.

Bandwidth

The key to equipment and the cable that interconnects it is bandwidth. If you need only 158 or fewer channels, that is 1 GHz of bandwidth. If you are delivering digital signals, you will want to look at the third harmonic. Channels 42 through 116 are commonly reserved for digital signals (Tables 7-2 and 7-3). Channel 116 ends at 750 MHz. The third harmonic of 750 MHz is 2.25 GHz. For digital applications, get cable tested and verified to 2.25 GHz. This extended bandwidth also applies to satellite-delivered programs that are also digital.

If other channel blocks are assigned by the FCC, further testing and verification of cable and equipment will be required up to the third harmonic of the highest frequency delivered.

The Future of Cable

Since the SCTE changed its T from television to telecommunications, it was obvious that the cable industry intended to deliver a whole lot more than just television. Right now the hot item is data delivery, but they're also talking about telephone delivery, even voice over IP, by running telephone lines as part of the Internet feed. The big problem cable has is reliability.

Right now, when there is a power failure, your telephone is backed up by massive battery installations in the phone company. Your Internet connection has no such guarantee. What if you want to call 911? Telephones are required to meet 99.9999 percent reliability, or 53 minutes of downtime per year. Cable currently averages 123 minutes of downtime each year, more than double the requirement.

8

Hiring, Buying, and Selling

In any complex technology, the simpler you can design it and install it, often the better it will work. Simplicity requires knowledge. The less you understand something, the more complex it will seem. Arthur C. Clarke, the noted science fiction author, has said, "Any sufficiently advanced technology is indistinguishable from magic."

KEEP IT SIMPLE

When NASA first started sending up astronauts, they quickly discovered that ballpoint pens would not work in zero gravity. To combat the problem, NASA scientists spent a decade and millions of dollars to develop a pen that would write in zero gravity, upside down, underwater, on almost any surface, and at temperatures ranging from below freezing to 300°C. The Russians used pencils.

Hopefully, the technology you will be dealing with will not seem like magic. It will make sense to you. Your job is to have it make sense to everyone around you. The more each project makes sense, that every screw and cable and machine has an obvious and simple purpose, the less complex the installation will be.

The beginning of that simplicity is understanding the process of purchasing the parts and equipment. This chapter is aimed specifically at how to buy. Just like your end user who is your customer, you are a customer of others. And just as your customer should be educated and aware of what you will be doing for him, you should be aware of what vendors, dealers, and distributors will do for you. If they did not have a purpose, they would not exist.

An Informed Customer

The purpose of your relationship with a dealer or distributor, indeed the purpose of this book, is to make you an informed customer. If you make decisions without the appropriate knowledge, you are an uninformed customer and will waste time and money. Once you are an informed customer, you can base your decisions on the facts, opinions, and influences around you. If you wish to buy speaker cable at $100 per foot, you will do it with full knowledge instead of faith or belief.

Table 8-1 list the three most important parts to any installation.

TABLE 8-1   What Is Important?

1.	Planning
2.	Planning
3.	Planning

You cannot plan too much. Nothing will speed up the process, simplify the work, save you money, get you on budget and in on time, and allow you to sleep at night as good planning. Planning is the virtual installation, the installation in your brain, and the brains of those around you, before the work begins. Planning allows you to anticipate changes, errors, or mistakes and make space for their solutions.

In fact, you really need to plan the plan. Because nobody can predict what your next project will be, how big it will be, or the details involved, what you need is a master plan that will work with any project, and can be expanded or reduced to fit the requirements. This planning will extend from this chapter to the installation chapter.

Wire and Cable Planning

Too often a designer or installer will wax poetic about some huge job she had designed or installed. And yet, have you ever noticed how often the choice of wire and cable is mentioned? Often, even the manufacturer is not mentioned, much less the part numbers chosen.

Even though the majority of work in an installation often concerns the wire and cable portion, it is the least considered. The fact that you bought this book, and have read it to this point, means that you realize the critical nature of the wire and cable decisions. I hope you now understand that the wire and cable choices are just as critical as any piece of hardware or software that will use it. You can ruin a perfectly good design or installation by a bad choice of cable, especially when running critical signals such as HD video.

Your digital installation is a complex digital network feeding critical data streams from point to point or even on a network architecture. Everything is important, possibly even critical, including your wire and cable choices.

Table 8-2 shows the most basic steps in any project, big or small. Each item on the list is phrased as a question. If you cannot answer "yes" to every question, you are not ready to start this project.

TABLE 8-2   Planning the Plan

1.	Do you know how big the project is?
2.	Have you actually talked to the end user/customer?
3.	Does the customer understand what you are about to do?
4.	Do you know how many people will be required to complete it?
5.	Do you have the crew and crew skills to complete the project?
6.	Do you know how many days will be required to compete it?
7.	Do you know what day the project will begin?
8.	Do you know what day the project will end?
9.	Do you have a schedule for each employee?
10.	Do you know what parts and equipment you will need to buy?
11	Do you know where you will buy all the parts and equipment?
12.	Do you know if you will "pre-fab" some or all of this project?
13.	Do you have plans and/or blueprints of the site?
14.	Have you added time and money for surprises?
15.	Do you know how much this project will cost you?
16.	Have you agreed on a price with the customer?
17.	Do you know how much profit you will make on this project?

$urpri$e$

There are surprises in every project. Surprises, as the title suggests, can cost you money. A surprise is something you did not know and weren't expecting. It can be the injury of a key employee, a financial reversal of the customer, forcing the closure of the job, or the unavailability of an important piece of equipment.

However, you can plan for each of these contingencies and many others. Table 8-3 lists many common surprises and what you can do to outwit them. The point is, you can plan for any contingency. As long as you have a plan, you can minimize the $ in $urprise.

TABLE 8-3   Surprises

Problem	Plan Ahead
Key employee injured, maybe even you!	Cross-train employees so that there is no single person for each task
Customer can't pay	Ask for financial information up front; run a D&B; pass on projects with iffy financing
Can't buy equipment/parts	Alternate manufacturers, alternate distributors
Staff argues about installation	Meet with the installation crew and discuss details on a regular basis; discuss the big picture and the little picture details

You can tell from Table 8-3 that any problem has a solution. Thinking is a whole lot cheaper than fixing, so plan for these contingencies. There are a million stories of installations that ended in disaster.

> The contractor bid on the plans for a one floor installation and didn't notice the note on the plans that said "9 nine more identical floors."
>
> The customer really had no idea what was going to happen, how long it would take, or how it would affect his house. Now you're cutting the project in half with an impossible deadline of next week.

The solution to these and any other problems is planning, planning, planning.

The Book of Contingencies

Buy a loose-leaf three-ring binder and insert some lined punched paper. Title it the *Book of Contingencies*. On each page write something that could go wrong at the top. Below that write every solution you can think of. Put these pages in order. The ones that will happen at the beginning of the project go first, the ones that happen at the end of the project go last. Or you can put them in alphabetical order by key word. Just put them in whatever order is easiest for you to find each page.

You may not have to refer to this book during a job. At the end of each job, add the problems that occurred but were not anticipated.

When disaster strikes, your brain can lock up. Go to the book and look up the appropriate section. If it's a new disaster, put it in the book and leave the page blank. When a solution is found, write it in the book before you forget.

Problem Solving

Here's an amazingly simple way to solve a problem. This works no matter how big or complicated the problem seems. Table 8-4 shows the steps. These steps may seem overly simplistic, but you will be amazed how well this system works. When your staff votes for the causes of the problem, they get to vote for any three. If you want to go even deeper, give them one vote each, have them vote on the three top causes, and put those three in order.

TABLE 8-4   Problem Solving

1.	Obtain a flip-chart and some thick pens
2.	Assemble the entire staff for the job, maybe even on site
3.	Tell them your concern; try not to say "this is the problem"
4.	Ask them what the problem is
5.	Write down exactly what they say
6.	Ask them about possible causes for those problems
7.	Write down every reason, no matter how silly
8.	Ask them to vote for the top three causes
9.	Ask them to suggest solutions for each of the top three causes
10.	Implement the solutions, one at a time

You might find that you and your staff are always complaining about the same thing. This repetition balloons a simple problem into a major one. Writing it down defines the problem, giving you amazing clarity. Often the fix is immediate and obvious. Having your entire staff present makes them part of the solution, and it makes them partly responsible for fixing the problem.

How to Say No

In any industry, the most difficult thing is to turn away business. However, if you ask some of the most successful installers or system integrators, they will exhibit almost a sense of pride in talking about the jobs they turned down.

Some installers turn away the majority of clients who come to them. Then they can "cherry pick" the best projects. Of course, you have to establish a reputation for unmatched integrity, quality, and workmanship before you get to that level of business.

Table 8-5 shows some of the lines you can use to refuse a job. Look at these reasons and decide which could be most believable. For instance, for reasons 1 and 2, decide the dollar amount that would make them true. Of course, you can always change your answer later. At least you will have a line with which to start.

TABLE 8-5   How to Say No

1.	The project is too small
2.	The project is too big
3.	You have never done this kind of project before
4.	You don't have time to train your staff
5.	The customer has a history of "changing his mind"
6.	The project will take more staff than you can recruit
7	The project is scheduled for an impossibly short period
8.	The project is spread over too long a period
9.	The project is geographically too far away
10.	Everybody and his brother will be bidding on this one
11.	You cannot make a reasonable profit
12.	You have too many projects already

How do you get to the top? Make fewer mistakes than your competition. Note that I didn't say, "make no mistakes." Nobody can do that. The trick is to learn from them and not repeat them. The company that posted the sign "Failure is not an option" had scared employees who took the tried and true path. Nothing new was ever tried. Nobody stepped out of the box to solve a problem. This is a recipe for failure.

What it should say is "Failure is *always* an option. Learn from it." You can use the good old suggestion box. It allows everyone from the janitor to the installation supervisor to have input. Good ideas very often come out of bad ideas. They don't come out of nowhere!

If you make fewer mistakes than the competition, in short order you will be able to use reason 12 in Table 8-5, which is the best one to have. We should all have such a problem!

Knowledge is power, so let's examine those questions in Table 8-2.

Do You Know How Big the Project Is?

If you don't know how big the job is, how will you decide how many people will be required to complete it? You can start by the dollar size or the bid size. You can also estimate the number of pieces of equipment or the number of rooms. You can even have a scale to determine the size of a project. If you are using a scale of 1 to 100, you, as a larger installer, could pass on the projects between 1 and 20; if you are a smaller installer, you might pass on any project above 70.

If you are a designer/consultant, it would be very helpful if you could quantify the job so that you know which installers could actually do the job. There is nothing more frustrating than an installer or contractor trying desperately to pull off a job that is way too big for it.

Have You Talked to the Customer?

Communication is the key to planning. If you are doing any installation, especially home installations, talking to the customer is critical. Most consumers are only vaguely aware of the technological choices available to them, and many have been confused by the claims made by high-end manufacturers and dealers. Be sure you understand what the customer expects, and what he expects from you and your crew, before the project starts.

Professional customers are better informed and have a better knowledge base than untrained consumers. However, many competent engineers are still using ideas and approaches that are decades old. Although it's not your job to 're-train' them, it is your job to advise them of their current options and prevent them from ending up in a technological dead end.

You can save them money, and move them into the future. If they see you as the source for answers, who will they come to next time they need a new edit suite or other expansion?

What Crew Size Will Be Required?

This decision directly affects your bottom line. Perhaps no decision is more critical. Pick a low number to keep the cost down and you have a crew rushing to make the deadline without enough hands to do the work. This leads to poor workmanship, with expensive reinstallations.

Too large a crew is a waste of money. They will be sitting around waiting for something to do. If you have to advertise for a crew, be sure and list the *minimum* skill set required in the ad. Whatever they don't know, you will be training them, or paying some one to train them for you.

Do You Have the Crew and Crew Skills?

Because you do not make the equipment and don't manufacture the wire and cable, the only thing you are selling is service. In that regard, the choice of staff and the skills they possess are really all you have to offer. Table 8-6 lists some suggestions to make your crew second to none.

TABLE 8-6   Crew Skills

1.	Upon hire, give each installer a hands-on test.
2.	Grade his or her work.
3.	Give them a rating for expertise.
4.	Use this rating to determine who can sub for whom.
5.	Offer paid training for skill improvement.
6.	Get manufacturers, distributors, dealers to do some training.
7.	Videotape your training classes for future new employees.
8.	Allow re-testing on a regular basis to upgrade standing.

I know one installer who listed every connector installed on a large board. Each connector showed every step in its installation, what cable to use, with manufacturer and part number, how to prep the cable, how to attach parts, boots, and strain reliefs. It even show what kind of heat shrink, size, color they preferred. That way there was absolutely no question as to how they wished their cables to be assembled. A quick look at the board showed everything. No employee could ever say "I didn't know" when they incorrectly wired up something. This attention to detail paid off handsomely in quality and consistency, not to mention profit.

How Many Days to Complete the Project?

There are many excellent software packages for project control. They allow specific timelines for each step of the installation. If you are wiring while a building is being built, often you can break up the process into discrete parts, such as a pre-wire after the studs are placed but before the walls are completed.

Sometimes you can subcontract this pre-wire to the electricians. However, you might want to be sure they understand the nature of the cables they are pulling, especially if this is for digital audio or video, and the limitations inherent on these cables. Of course, it is unlikely you'll be around to see them pull it in, so you might want your crew to do it instead.

What Day Will the Project Begin and End?

You and your entire staff, including clerical staff, should be very aware of start and end dates. This is doubly true if you have penalties that kick in if you go beyond the end date. This is where you should definitely add some surprise time.

Even a day or two extra is a godsend. If you don't use those extra days and come in early, you'll look like a genius.

A Schedule for Each Employee?

This goes hand in hand with crew size. If you have an accurate time plot of what needs to be done and when, the number of people to do it will be easy to calculate. If many of your installers are private contractors, be aware that, when times are good, they may not be available. And the better they are, the less likely it is that they will be available.

Talk to employees and determine what they consider to be a fair day's work. Overburdened installers are unhappy employees. Find a reasonable compromise. After all, you will be asking them to do their best work, and keep it up day after day.

You should post the work schedule at the office and at the job site. Together with the build timeline, this will allow the other contractors to see when your crew will and won't be there, giving them no surprises. It will also allow the end user/customer to see your progress, so he will not bother you unnecessarily, especially during the crunch time at the end of the project. Contractors and customers also will know from the posted schedule when any particular member of your staff will be on site.

Customer Liaison

It might be a good idea to assign the job of customer liaison to one of your crew chiefs, supervisors, or other knowledgeable person. Make sure this member of your staff knows about the job in detail and has good interpersonal people skills. The person can interpret what is happening to the customer and translate suggestions or changes from the customer to you and your crew. This customer liaison would be an excellent choice to sub for you when you are not available.

What Parts to Buy and Where to Buy Them

Do you know exactly what you will be buying for this project? And do you know the companies from whom you will purchase these parts? Do you have an account with them? Cash up-front or on-delivery can cause unnecessary frustration. Open accounts with every supplier. For each source, have an alternate source.

This applies to passive components, such as wire and cable, as well as the equipment itself. What if a manufacturer discontinues a product you have specified just as your project begins? You should have an alternate part number from the same manufacturer, or possibly a different manufacturer altogether. What if a distributor goes under during your project? An alternate distributor or dealer should be in your *Book of Contingencies*.

Dealers and Distributors

Most electronic parts and equipment are bought from distributors. From here on, I will use the word *distributor* but it could be a company, or even a single person, with a relationship to a manufacturer. It could be a "box house" that inventories equipment but nothing else. It could be a dealer that carries major and minor support gear. Or it could be a distributor that handles a wide range of parts and supplies. Sometimes other contractors or system integrators have such a relationship and can supply items and integrate them for you.

Most of the time, you or your purchasing department are looking for someone to supply as many of the parts and equipment as possible—a "one-stop shop." Just be aware that, if you are in a hurry to line up thousands of items, major financial mistakes can be made. Table 8-7 lists the 29 basic questions that you or your purchasing department should ask of any distributor.

TABLE 8-7   Distributor Questions

1.	Do we have an account with you?
2.	Is our account in good standing? Is it an open account?
3.	What are the restrictions, if any, on our account?
4.	What pricing do we get? Do we get your best pricing?
5.	What do we have to do to get your best pricing?
6.	How many lines do you carry?
7.	Do you carry the lines that I need?
8.	Where do the lines I need stand in your top-lines list?
9.	Are you an authorized dealer/distributor for the lines I need?
10.	Are you a "specialist" in the products I use?
11.	Do you get "specialist" pricing from the manufacturer?
12.	Do you stock the lines I use?
13.	Do you stock the part numbers I use?
14.	Where is this stock physically located?
15.	If stock is not local, what are the additional shipping costs?
16.	If needed right away, what is the additional shipping cost?
17.	Do I have a dedicated inside salesperson assigned to me?
18.	Do I have a dedicated outside salesperson assigned to me?
19.	Do you have other branches in other locations if I need them?
20.	Do I have the same availability of lines and products there?
21.	Do I have the same pricing there?
22.	Will you give me contract pricing or another price guarantee?
23.	Will you sequester stock? What are the requirements?
24.	Can you provide special constructions?
25.	Will you inventory special constructions for me?
26.	Can you provide just-in-time delivery?
27.	Can you stage deliveries?
28.	Can you stage deliveries to multiple sites?
29.	Are there additional charges for staging? What are they?

Some of these questions seem obvious or even simplistic. If so, just imagine if the answer to some of these obvious or simple questions were "no," or the distributor couldn't answer the question. What would you do? Your *Book of Contingencies* should have the answer. The details behind each question follow.

Do we have an account with you?

Of course, you should know the answer to this question. If you don't, you are a very small fish in their pond or you'd better correct their misconception. If they don't know that you have an account, what kind of pricing are you to expect? List plus a lot, one would imagine.

Is our account in good standing?

Your accountant told you he was delaying payment to this distributor for 90 days, but you forgot about it. Now the account is closed, or possibly cash on delivery only. And they are the key source for the parts you need. Lucky you asked this "obvious" question. Of course, you have an alternate in your *Book of Contingencies*. Hopefully, the bill for that contingency distributor was paid on time. Maybe you should add some rules about bill-paying to your book.

What are the restrictions on our account?

Especially on old accounts, set up long ago by people long gone from your company, you may find that you have a restricted account. It may be restricted to certain lines or a certain dollar amount. Sometimes they don't want to tell you what the restrictions are, but you'll find out the hard way when you order something not on the "approved" list, or you need that big ticket item but it is over your limit. So find out before you start to order parts or equipment.

What pricing do we get?

This may be the hardest question to get answered. Sure, they'll tell you what your *price* is. They just won't tell you how it compares to the prices of other contractors, installers, or system integrators. This doesn't prevent you from asking hard questions. A dealer may tell you, "We charge everyone the same amount for the same quantity. If you buy more, we'll charge less per item." If you find out this is not true, bring it to his attention. Simply ask him what you need to do to get to that next level of pricing. It is now up to the dealer to reveal just what the levels are, and how you can reach them.

Of course, you may not have the volume to reach those new levels, but at least you will know what they are, and have a goal to reach those levels. Alternate suppliers may set the lines lower to entice you to change. Just be sure they have the inventory and relationships to support you adequately.

What do we do to get best pricing?

If your company is growing, and you can show this to a distributor, you can make a good case for getting preferential pricing. Yes, this is a lot of work but you can potentially saves thousands, even millions, of dollars by having "relationship pricing." What the distributor loses in margin, you will make up in volume. Of course, you will have to guarantee levels of purchasing while your company grows, so there is some risk involved.

In some cases, the distributor may require you to commit to a certain amount of purchasing per year, per quarter, or other period. If you don't make that level, you are required to buy the amount agreed. This can get you some really aggressive prices because the distributor is now "guaranteed" your business.

It is illegal in the United States for a manufacturer to set pricing for a distributor. A distributor who tells you that he can sell only something at list price is not telling the truth. He could even sell you something at his cost, or even less. Of course, this would be likely only if the distributor is guaranteed to get your business on other items. Then the free items become a "loss leader" to entice you into a "package deal."

The problem with a package deal is that it becomes very hard to tell what you are paying for any particular item. Sure, the total looks good, but you will have a hard time comparing it to line-item pricing from someone else.

How many lines do you carry?

This question can have some major repercussions. If this is a small distributor and they carry 500 lines, how much attention, much less inventory and warehouse space, can they give to each line? It is more likely that 20 or 30 lines account for the majority of their business.

Companies that specialize in just a few products or types of products indicate a more focused approach. For instance, a distributor might specialize in wire and cable products. Around those, I would expect some connector lines, wire ties and ducting, maybe trays and ladders, and the tools to install them. They might have serious depth to their inventory and have stock in almost every variation you might need. They got that depth by sacrificing breadth or a large number of product lines. Maybe a "one-stop shop" isn't all it's cracked up to be!

Where are my lines in your top-lines list?

Ask them if they have a list of lines by sales. You don't need the dollar amounts, which they won't show you, just the order the lines are in. If the lines you will need are not in their top 20 or 30, you should ask some pointed questions about supporting you projects.

If you're a good customer, they may offer to bring into stock the specific parts you need from those other lines. But having them inventory specific parts for you is costly. They are much less likely to keep their inventory up when it falls low. And if you change your mind and need a different part, do you think they will? You'll never hear the end of those parts you said you'd need but didn't buy.

If the distributor carries that line in considerable breadth and depth, they probably already carry all the parts you will need. Then you are not the only customer buying them, so they wouldn't be doing you a special favor by putting them in stock.

Since they inventory those parts in greater breadth and depth, their technical knowledge will be greater, too, and they might help you avoid a bad decision. If all the other installers they service buy part X and you're ordering part Y, they should at least tell you that you're buying something out of the ordinary. Then you can make an informed decision.

Also, the technical knowledge of the distributor's sales staff cannot go much further than the top 20 or 30 lines. After all, a manufacturer's sales people promote only one line, and many of them aren't knowledgeable about even that line. How could a distributor sales person know 20 or 30 times as much? Therefore, if it is a line they carry specifically for you, don't expect much in the way of technical assistance besides referring to the catalog. You can do that yourself.

Manufacturer technical support

Technical support makes it essential, even vital, for you to have contacts with the manufacturers of the parts and equipment you buy. Believe me, they're happy you are buying their parts or equipment through a distributor. They would be more than willing to back up your purchase with technical assistance.

You might want to back up those names and phone numbers with a web address or specific e-mail address. When you have a problem at 2 A.M., it is unlikely that any manufacturer will have a technical person available to help, unless you are buying parts and equipment from Europe. If you are calling Europe you'd better have the name of a technical person who speaks English, unless you are multilingual.

Languages

While we're on the subject of languages, keep track of your employees and the languages they speak. Having someone on staff who speaks Spanish or some other language can be a godsend. What if the other contractor working at the site has workers who speak Spanish and very little English? Your Spanish-speaking employee can avoid a lot of headaches by translating your instructions.

If you're buying parts from Spain or Mexico, for instance, that same employee can speed up the process for you. If there are problems, he can help solve them by eliminating any language barrier.

When you hire employees be sure and note what other languages they speak. Those who speak Chinese, Japanese, Russian, Polish, Swedish could be valuable to you. They could even lead you to customers who speak that language better than English.

Are you an authorized distributor?

This has got to be one of the most surprising questions, mainly because of the answers you will get. Of course, you want to hear "yes." Instead you may hear,

> We have access to that line.
> We can get that for you.
> We distribute that line.

...or variation on that theme. If you get an answer besides "yes," you might want to contact the manufacturer and ask for a list of authorized distributors. Sometimes they are listed on the manufacturer's web page.

Redistribution

Many distributors are not directly connected to the manufacturer and have access through another distributor. Some manufacturers have re-distributors that are given special pricing specifically to service other distributors. Of course, those distributors are not specialists and will have less knowledge of a specific line than a true authorized distributor.

YOUR TAX DOLLARS AT WORK

I once met a distributor who had no authorized lines. He used to work as a government buyer so he knew the ins and outs of government contracts. He would get redistribution pricing on everything, double that, and submit a bid. Because most other distributors didn't even know there was a government bid, much

less how to read or submit a bid, he would win every
tenth or twentieth bid. He made millions and millions
of dollars, from a one-man office, without every having any inventory or carrying a single line.

If the items in question are minor to your installation, for
example, a bag of screws or some wire ties, perhaps this is a
good thing because you can have more of a one-stop-shop
that will simplify your purchasing. Just be aware that you
will pay for this simplicity.

Price will let you know if you're dealing with a redistrib-
utor. It is very rare that his price will be the same as an
authorized distributor. If the price on certain lines is consistently high in his bid to you, it is quite likely that he is a re-
distributor. Distributors love to say, "Sure, we can throw
that in." They then go to the authorized distributor and
make a deal to include that item in their offering.

Returns

Redistribution, or the use of nonauthorized distributors, can
get really messy if you have a return. What if the color is
wrong, or the wrong part was ordered, or someone substituted something you never asked for? Having a return like
that is akin to putting the toothpaste back in the tube. A
direct authorized distributor will cause fewer problems.

Returns might be a good section for your *Book of
Contingencies*. What is your distributor's or the manufacturer's return policy? Items with a restrictive return policy
better be on your list of "parts we use all the time." That
way, you're more likely to use them on the next project
instead of just throwing them away.

Custom-made items are particularly difficult to return.
For instance, if the cable you order is special it is probably

nonreturnable. After all, who is the distributor or manufacturer going to sell it to? Therefore, you should calculate the length needed with the greatest accuracy. Of course, if you are ordering a small amount, there is probably a minimum from the manufacturer. You will have to charge the entire cost to the customer, and throw away the excess or create a design that you can use on future jobs.

Custom-made cables

While we're talking about specials, there are many companies who have almost all their cables custom made. That way, they can put their own company name on the cable, with their web site or 800 number and their company color. The cable itself may even be color-matched to the company color.

You have the option of including the manufacturer's name or omitting it all together. Adding the manufacturer' name can be an indication of the quality of the construction. Most savvy customers will know that you are not a manufacturer and didn't make the cable yourself. Omitting the manufacturer's name means the customer can't go around you and find an equivalent in the manufacturer's regular line. The customer will have to come back to you for more cable. Then it's a question of, do you want to become a distributor as well as an installer?

Is the distributor a specialist?

This goes back to the number of lines a distributor might carry. You know what they say, "Jack of all trades, master of none." This applies to distributors, too. If they carry a large number of lines, it is doubtful they could afford the time to specialize in any of them.

If the items you are buying from them are complex and expensive, it is imperative that they be specialists. If they are not, then you will heading for the manufacturer when you have a problem. Don't expect the distributor to be any help. If you got the deal of the century from a distributor, that also means he made so little profit off you that he cannot afford to help you. His service ended when you picked up the box.

Of course, you can always ask if the distributor is a specialist in a particular line or type of product. Some manufacturers list their specialists amongst their general-line distributors.

Do you get special manufacturer pricing?

In some cases, these specialists might get preferential pricing from the manufacturer for focusing on a particular part of that manufacturer's line. Ask them. If you're still not sure, ask the manufacturer.

Do you stock the lines I use?

Certainly, this is a basic question. If you have a list of parts and equipment you will use for a project, e-mail or fax this list to a distributor. Your intention is not to get pricing, but to find out if they carry these parts.

Allow them to substitute or suggest other manufacturers. Then you will know where their strengths lie. You will also know where their expertise lies. If they are a wire-and-cable specialist, don't expect much in test equipment, unless it is related to wire and cable.

If they cover only a few lines that you need, don't dismiss them. Add this information to your *Book of Contingencies*. You will never know when the other distributor is unavailable and this small focused distributor will save the day.

Do you stock the part numbers I use?

Just because they carry the line you want, don't assume they carry the parts you need. For instance, a wire-and-cable specialist may focus on data products, or industrial cable. Calling him for audio or video cable would be a waste of time.

That's why the list you send them should include every detail, every part, description, and part number. If you have misgivings, you could always ask them to indicate what they have in inventory on each part number. If you're going to use 20,000 ft of precision video cable, and they have 500 ft in stock, that pretty much says it all.

Where is this stock physically located?

Ask this of a few very large distributors and their answer might be, "Oh, don't worry, we can have it overnight to you." That may be fine for some jobs, especially if you are ordering large quantities. Of course, if you are in New York or San Francisco and the stock is in Chicago, you're not going to be jumping into your pickup to pick up a roll or two in an emergency.

You really have to weigh convenience against price. Some distributors with large central warehouses can give you excellent volume pricing.

The tail wags the dog

If you are a very large contractor, installer, or system integrator or you have a very large project, there is nothing stopping you from talking to the manufacturer. Often you and the manufacturer can come up with an arrangement and then you can shop around for a distributor that meets your requirements and is willing to take this business.

Of course, everything will be on the table, including the margin to the distributor, so it's a matter of take it or leave it. This doesn't prevent the distributor from requiring some guarantees as to volume, time, warehousing and so forth, to protect its margin. Everybody deserves to make a profit and stay alive, including you!

If stock is not local...

The first question should be, where is it? Do you have the part numbers I need? What inventory do you have on these part numbers? In the case of wire and cable, don't forget to specify colors, if that is what you need.

Sometimes they will throw in the shipping. Just be sure this is spelled out in writing before you commit to the deal.

The next set of questions is, How can you get this to me? Truck? Air? What will the additional expense be? You can always specify a carrier. Sometimes it is cheaper for them to use your overnight account for emergency shipments, than to use the distributors and have them charge you.

Dedicated inside salesperson

There is nothing more frustrating than calling your distributor in the middle of a large project and having the salesperson not know who you are or anything about your project. Have the distributor assign you an inside salesperson right from the start who can help you when it comes to crunch time.

Dedicated outside salesperson

While you're at it, ask for a dedicated outside salesperson. If you are in a busy period, you might sit down with the outside salesperson and arrange a schedule for him or her to stop by. If you have a list of parts generated for each job or just a general boiler plate list, this is a good time to pull it out. Give the salesperson a copy. Have the salesperson go back to his or her office and determine which items the company carries, and in what depth and breadth, and return it to you. Be sure he or she shares the list with the inside salesperson so that all three of you are ready. Maybe you can take them both to lunch.

Creating a relationship with these people is worth its weight in gold. When you're desperate for a roll of wire or bag of connectors, who will deliver it to you? The distributor who knows you as a name on a screen or the one you took to lunch? Make sure they understand that they are critical to your success. Have them contribute to the success of your company.

Other branches in other locations

If you intend to do projects outside your immediate geography, you have to ask yourself, How will I supply all the parts and equipment for those jobs? Start by asking the distributor. Do they have other branches? Where are they located? Do I have the same availability of lines and products there? Do they have local inventory? Is it the same as the branch with which I am now dealing? Do I have the same pricing there?

If you intend to do many jobs in that location, maybe they can assign you inside and outside salespersons from that branch. If they have one location, or they don't have a branch where you want to do business, ask them how they intend to service you. If they don't have a solid and convincing suggestion, perhaps you should start looking for a different distributor in that location.

Big distributor, small distributor

Here you have a quandary. If you pick a small distributor, you will be a big fish in their little pond. You will get better customer service. Of course, their inventory might be a little thin, but they will make up for it by helping you solve your problems. One thing you can do is provide them with a list of what you will need and make them guarantee their inventory levels on those items. There are many very large firms that buy from small distributors, especially if those distributors are focused on one specific part of the industry.

A big distributor, one with dozens or hundreds of branches, will have a huge inventory. Of course, it might not always be in the location in which your project is located, but they can get it to you. Their computer probably shows every branch, so with each item always ask, "and where is that part located?"

The problem is that your project is a blip on the screen for them. You'll know because your "assigned inside and outside salespeople" will seem like a revolving door. Of course, they will have almost every little thing you want. If they don't, it will be extremely difficult for you, now a small fish, to special order it.

Contract pricing or other price guarantee

One approach that is to be encouraged is to create relationship pricing with a distributor. If you can get him to maintain specific inventory levels, it is only one more step to guaranteed pricing. This arrangement could be for a specific project or a specific period, such as a year. They might require you to buy at certain levels per month, per year, or for each project. After a few years in business, you should have a very good idea of just how much business you can guarantee. Of course, you can always buy more than the agreed minimum and maybe extend the volume and pricing for the following year.

Meet comps and project pricing

If you do have major projects, be aware that many manufacturers like to lock in those projects with "meet comps." These are pricing arrangements where they will "meet the competition's" pricing. This arrangement is done by the distributor and can result is a significant price reduction to the distributor. In this case, you get the competitor's price, the distributor makes his margin, and the manufacturer takes the hit but is compensated by volume.

Of course, some distributors will "create" a meet comp out of thin air by mentioning a competitor when the business wasn't going that way at all. If discovered, and they often are, the manufacturer can take any one of a number of punitive steps, from barring further meet comps for some period to dropping that distributor for flagrant and repeated violations. The last thing you need is for a distributor to become unauthorized. Don't encourage false meet comps.

Some manufacturers use "project pricing" or some similar program. The distributor "registers" each project with the manufacturer. Then that distributor, and only that distributor will get special pricing for that project. The intent is that the distributor who did the work, found the project, and developed the relationship is rewarded with a rebate. He can keep that rebate or share it with you by offering better pricing. If your project is $250,000 or more, ask the distributor if he has registered your project.

Sequestered stock

Sequestered stock is stock held in the warehouse by the distributor and intended for a single customer. It may be a standard part number of a manufacturer but it will not be sold to anyone else. It is the ultimate in inventory control.

Of course, the distributor doesn't like it because it means less cash flow. You will have to make it very much worth his while to get such treatment. This is possible only with the largest jobs or the highest profile customers. You will know right away if you quality for either classification.

Special constructions

If you order specials, you probably will want the distributor to stock them for you. Unless you have considerable warehouse space, the distributor is the best place for it to stay. Distributors will resist because, like a manufacturer, they get paid when the item is shipped to you and you pay for it. If it is sitting on their shelf, they're not being paid. Further, they are taxed by the inventory on their shelves. Including your specials. Therefore, the price for this service will be considerably more than if you just take it. The tax burden will then be yours, of course.

Just-in-time delivery

The best for you and the worst for the distributor is *just-in-time* (or JIT) delivery. This involves a schedule that you develop with a window in which items are to arrive—not before, not after. This is the hardest thing for a distributor to do, and he will charge for this. It also means he must keep special inventory because of the penalties if he doesn't deliver.

Most commonly, this type of delivery applies to very large original equipment manufacturers (OEMs), but also can be used on large projects with limited warehouse and storage space such a major office buildings, stadiums, and theme parks.

Can you stage deliveries?

Stage deliveries are a little less intense than JIT. For instance, every 1,000 ft of cable will include 20 connectors and 20 wire ties, or whatever. The point is to have all the parts assembled in one package. You will pay for this, but it can pay off in reduced and simplified labor costs. Will it be cheaper for you to do the staging or have the distributor do it instead?

Can you stage deliveries to multiple sites?

If you have a project that involves numerous sites, staging each site can save you a lot of hassle. This means that everything you need, or at least everything the distributor carries that you need, will arrive in one big package.

If you have the warehouse space and someone to do the staging, it is probably cheaper for you to do it yourself. If you don't have the space, or the staff, a distributor is a good second choice.

The real question is, What will this cost me? Anything that involves counting, packaging, and distributing will be expensive. Find out how much more it will cost so that you can make an appropriate decision.

Dealing direct

Some manufacturers may lure you with the offer of "going direct" to them. This means that you go around the middleman, the distributor, with the promise of a significant cost saving. Sometimes this promise is kept, sometimes it is not.

If you are offered a "direct" relationship with a manufacturer, get your pricing from them and compare it to your distributor's pricing. Tell your distributor the volume you will do, if he doesn't already know, and give him the direct prices as your target pricing. Many times, I have seen distributors beat direct manufacturer's pricing. So what other advantages are there to dealing directly with a manufacturer?

Will they keep inventory for you? Manufacturers are now taxed on inventory, so it is in their best interest to have as close to zero inventory as possible. In their ideal world, you order something, they make it and send it to you. If you're lucky, they make what you want as you order it. If you're not lucky, you're on a schedule.

If that schedule is 5 weeks or 16 weeks, how will that affect your project? What happened to the stuff the manufacturer made last week? That is now on the distributor's shelf. In other words, where manufacturers use distributors, those parts are on the distributors shelves. Those distributor shelves are the warehouse for the manufacturer.

If you get involved with special constructions, which are handled by a distributor but made by the manufacturer, you will learn all about schedules and time tables. If you are contemplating having cables made for you, you should get an estimate on construction long before begin your project. The last thing you want is a specific piece of equipment or a certain cable holding up the completion of an installation.

On site or "pre-fab"?

There are two ways to build an installation. The first is on site, at the customer's location. The second is to prefabricate (pre-fab), some or all of the project. Hopefully, you know which you will do and to what extent. We will discuss this in much greater detail in the next chapter.

Plans and/or blueprints

It would be hard to estimate the size of a job without plans. A job small enough to not require blueprints or line drawings is a very small one. Unless you are doing home installations, in which case plans may be optional, plans can be an excellent aid to control the process. Plans can help you perform the tasks listed in Table 8-10. This is a simple list, and we'll go into much more detail in the next chapter.

TABLE 8-10   Plan Tasks

1.	Number of racks or other equipment containers
2.	Floor space/footprint of racks and containers
3.	Equipment in each rack or container
4.	Cables within racks between equipment
5.	Cables between racks
6.	Cable type
7.	Cable length
8.	Total of each cable required
9.	Connectors required
10.	Conduit, tray, cable management length and type (if any)
11.	External wiring, power, speakers, monitors, and so forth

Time and money for surprises

In an ideal world, you could add a week of extra time to each project, and 20 percent on the budget, and still be the winning bid. Unfortunately, it is just those items that are sacrificed to make a winning bid. Add as much as you can.

How much will this project cost you?

By the time this process is finished, before the installation
begins, you should know, within a few percentage points,
just what this project will cost you. It is then a simple matter, based on the bid price you have given the customer, to
determine your profit margin.

Installation companies that routinely bid in single digits,
below 10 percent, do so simply to get the business and keep
some cash flow going. This can be very dangerous. One day
over or a few pieces of equipment mispriced will wipe out
that 10 percent. You are then losing money.

A top-of-the-line installer or system integrator can make
50 percent or more, some make over 100 percent. The minimum survival profit is somewhere around 30 percent, and
this is based on the entire cost, including labor, parts and
equipment, shipping, transportation, travel and hotel
expenses, warehousing and storage—everything. Add your
margin to the *total* cost.

9

The Installation

Putting It All Together

We've finally come to the installation chapter. This is where we stop asking *why* and start asking *what* and *how*. If you want to know why, you can find it in the previous chapters. The proof is in the pudding, as they say, and the proof of a good installation is in the performance of that installation, its robustness, and in its simplicity.

Simplicity is at the core of this chapter. In fact, you might find it a little too simple. Truth is, once you know why, the what and how is equally simple.

Subcontractors

The easiest way to get something done is to have someone else do it. That's an easy way, not a cheap way. Still, if there is some expertise required, it is often easier and faster to use an expert.

I would make a list of experts in my *Book of Contingencies*. If you're up against crunch time, using a subcontractor can be one way to save a project from disaster. Of course, you will be trading your margin for getting the project done correctly and on time. If you are new and trying to prove yourself, this may not such a bad trade.

The next step would be to check into training you or your staff to do what the expert did. Then you could keep the margin for your company and advertise that expertise in the flyers and literature you make for your customers.

Code

It is assumed that you know the code requirements in your area. I don't mean the NEC or electrical codes. I mean building codes, safety codes, and dozens of other codes. If you are working in a new area, you should read up on the local codes. It may be different, sometimes very different, from the code with which you are familiar.

One solution is to align yourself with a person or firm who is versed in the code requirements for the geographic area in which the installations will take place. Just make sure that person works regularly in that area and is very familiar with the specific code.

Another solution may be to hire a contractor or subcontractor who works in the area. For digital upgrades, this is often the solution. If you don't have the staff and or the time, then hire someone who does. It's still your project, so you still call the shots. You can use the following sections to ask the right questions and check that everything is being done appropriately, at least for the wire and cable portions.

NEC Considerations

If there is one error that can often escalate into disaster, it is ignoring the NEC. If it were federal law, things would be a lot easier. Because it is a voluntary code, some areas can adopt it and others not. There may be areas, such as Las Vegas or Chicago, that feel the NEC is not strong enough and have created an even stronger code.

THE NATIONAL ELECTRICAL CODE IS PRODUCED BY:

National Fire Protection Association (NFPA)
1 Batterymarch Park
P.O. Box 9101
Quincy, MA 02269-9101
Main Switchboard: 617-770-3000
Fax: 617-770-0700
Orders: 800-344-3555 or 617-770-3000

How do you know what applies in each area? Even when you know that the city or county uses the NEC, what does this really mean? Like every law, the NEC is open to interpretation, and a building inspector or fire marshal may interpret something very different from you, your architect, or your contractor. So how do you find out?

Go to the planning commission. Go to the board of permit appeals. Ask them if they go by the NEC, and how they interpret the code. Table 9-1 lists just a few issues that could cause you heartache, if you don't really know what your local officials think.

TABLE 9-1   NEC Ratings

Subject	Controversy
Plenum space	Ask them to define what is and what isn't a plenum. Often they will say any hidden space is a plenum, even if it is not connected to ventilation or air conditioning. How must air be fed and removed from racks so that the areas inside or outside are not plenum? How must cables be fed into racks to maintain their non-plenum rating?
Plenum cable	Is a cable marked CMP, CL2P, or CL3P sufficient? Some inspectors have erroneously required "all Teflon®."

continued on next page

TABLE 9-1   NEC Ratings (continued)

Subject	Controversy
Riser	Ask them to define what is and what isn't a riser. Is an elevator shaft a riser? Some erroneously assume that using riser-rated cables makes firestops unnecessary.
Other ratings	CM, CL2, CL3—when and where are these acceptable? Be sure your interpretation agrees with theirs.
Conduit	Does every cable installed have to have some rating? Can you put unrated cable in conduit?

Installing Each Cable Type

The following tables describe each cable type. Each section begins with a table that shows the limitations of each cable in basic parameters. A section on NEC ratings follows each, and a "helpful hints" section completes each one.

TABLE 9-2   Unbalanced Analog Audio

Cable type	Single conductor shielded
Gage size	No standards, bigger is more rugged
Signal type	Unbalanced
Cable impedance	Doesn't matter
Special requirements	Low capacitance is better
Maximum distance	Depends on system impedances
Rule-of-thumb distance	28 ft. with 10k Ω impedance
Minimum bend radius	Doesn't matter
Connector type	RCA most common
Connector impedance	Doesn't matter
Connector requirements	Indoor only
Other considerations	Use baseband video cable.
Other considerations	Use Category 5/5e/6 UTP with baluns
Cautions	Do not use CATV/broadband coax

TABLE 9-3   Unbalanced Analog Audio NEC Considerations

Unrated	Most common rating for home use
CM	Minimum suggested for commercial use
CMR	Available with precision video coaxes
CMP	Available with standard or precision video
Other considerations	Most runs too short to consider conduit

TABLE 9-4   Unbalanced Analog Audio Helpful Hints

1.	Keep interconnects short
2.	Keep connectors clean for good contact
3.	Using adapters to get to RCA will add to intermittent failures

TABLE 9-5   Balanced Analog Audio

Cable type	Twisted-pair shielded
Gage size	22–26 AWG; 24 AWG most common
Signal type	Balanced
Cable impedance	Doesn't matter
Special requirements	Low capacitance is much better
Maximum distance	Depends on system impedances
Rule-of-thumb distance	1,000 ft; more is quite feasible
Minimum bend radius	Don't break conductors or braid wires
Connector type	XLR, crunch blocks, many others
Connector impedance	Doesn't matter
Connector requirements	Indoor and outdoor available
Other considerations	Flexible braid/French Braid shields
Other considerations	Category 5/5e/6, no baluns needed
Cautions	Calculate maximum pull strength

TABLE 9-6   Balanced Analog Audio NEC Considerations

Unrated	No longer accepted by NEC for installations.
CM	Most common
CMR	Available
CMP	Available but expensive
Other Considerations	Conduit size based on cable size/fill ratio

TABLE 9-7   Balanced Analog Audio Helpful Hints

1.	Punching down wire is faster, gas tight, and reliable
2.	Put tubing or heat shrink on bare wires before punching down
3.	Smaller diameter means smaller gage, less ruggedness

TABLE 9-8   Speaker Cable

Cable type	Zip cord, jacketed twisted pair
Signal type	Unbalanced
Gage size	10 AWG at home, 12 AWG commercial
Cable impedance	Doesn't matter
Special requirements	Jacketed pair is more rugged
Maximum distance	Depends on resistance loss
Rule-of-thumb distance	250 ft; 12 AWG, 0.5-dB loss
Minimum bend radius	Doesn't matter
Connector type	Spade lugs, round lugs, bare wires
Connector impedance	Doesn't matter
Connector requirements	Indoor only
Other considerations	Woofer draws most current in bi- and tri-amps
Other Considerations	70-volt system allows smaller wire size

TABLE 9-9   Speaker Cable NEC Considerations

Unrated	Home installs, where visible
CM, CL2, CL3	Common
CMR	Not available
CMP	Available but expensive
Other considerations	Conduit

TABLE 9-10   Speaker Cable Helpful Hints

1.	Bigger is better than smaller
2.	If you really want to listen to speaker cable, go right ahead

TABLE 9-11   Unbalanced AES3-id or S/PDIF Digital Audio

Cable type	Coax
Signal type	Unbalanced
Gage size	Makes a difference at extreme distances
Cable impedance	75 Ω
Special requirements	The lower the capacitance, the better
Maximum distance	Depends on gage size, sampling rate
Rule-of-thumb distance	300 ft, 192-kHz precision RG-59
Minimum bend radius	10 × diameter
Connector type	RCA (S/PDIF), BNC (AES3-id)
Connector impedance	Doesn't matter, wavelengths are long
Connector requirements	Indoor only
Other considerations	Constant-impedance passive splitters
Other considerations	Can use Category 5/5e/6 with baluns
Cautions	No combining digital signals passively

TABLE 9-12   Unbalanced Digital Audio NEC Considerations

Unrated	Can be used for home where visible
CM	Commonly available
CMR	Most common in precision video coax
CMP	Available but expensive

TABLE 9-13   Unbalanced Digital Audio Helpful Hints

1.	Cable impedance is important
2.	Lower capacitance is very important
3.	In larger installations, keep aware of distance limitations

TABLE 9-14   Balanced AES3 Digital Audio

Cable type	Shielded twisted pair
Signal type	Balanced
Cable impedance	110 Ω ± 20 percent
Special requirements	Low capacitance
Maximum distance	Depends on AWG and sampling rate
Rule-of-thumb distance	350 ft, 24 AWG, at 192-kHz sampling
Minimum bend radius	10 × Diameter (affects impedance)
Connector type	XLR
Connector impedance	Doesn't matter
Connector requirements	Indoor/Outdoor versions
Other considerations	Category 5/5e/6, without baluns
Other Considerations	Impedance-specific splitters only
Cautions	Standard analog audio cable <50 ft

TABLE 9-15   Balanced Digital Audio NEC Considerations

Unrated	No longer accepted by NEC for installs
CM	Most common
CMR	Not commonly available
CMP	Available but expensive
Other considerations	Conduit size based on cable size/fill ratio

TABLE 9-16   Balanced Digital Audio Helpful Hints

1.	Multing and half-normaling patch panels not acceptable
2.	No passive combining of digital signals

TABLE 9-17   Analog Video

Cable type	Coaxial cable
Signal type	Unbalanced
Cable impedance	75 Ω
Special requirements	Good impedance tolerance, ±3Ω
Maximum distance	Depends on equalization
Rule-of-thumb distance	1,000 ft
Minimum bend radius	10 × diameter
Connector type	BNC; others acceptable
Connector impedance	Doesn't matter
Connector requirements	Indoor only; some outdoor versions
Other considerations	Precision cable, double braid
Cautions	Do not use CATV/broadband coax

TABLE 9-18 Analog Video NEC Considerations

Unrated	No longer accepted by NEC for installs
CM	Common
CMR	Becoming commonly available
CMP	Available but expensive
Other considerations	Conduit size based on cable size/fill ratio

TABLE 9-19 Analog Video Helpful Hints

1.	Determine pull strength in multicoax pulls
2.	Distance also influenced by size of center conductor

TABLE 9-20 Digital Video

Cable type	Coaxial cable
Signal type	Unbalanced
Cable impedance	75 Ω
Special requirements	Return loss < −20 dB to 2.25 GHz
Maximum distance	Depends on gage size and data rate
Rule-of-thumb distance	300 ft, precision RG-59 at HD
Minimum bend radius	10 × diameter
Connector type	BNC
Connector impedance	75 Ω to third harmonic (2.25 GHz)
Connector requirements	Indoor only
Other considerations	Wide range of cable sizes
Cautions	No SRL measurements acceptable

TABLE 9-21   Digital Video NEC Considerations

Unrated	No longer accepted by NEC for installs
CM	Common
CMR	Becoming commonly available
CMP	Available but expensive
Other considerations	Conduit size based on cable size/fill ratio

TABLE 9-22   Digital Video Helpful Hints

1.	Distances are halfway to cliff; read bit errors after halfway
2.	Everything you do is critical
3.	Beware of tight or regularly spaced wire ties; consider Velcro®
4.	BNC feedthroughs and barrels must be 75 Ω

TABLE 9-23   Machine Control

Cable type	Two twisted pairs individually shielded
Signal type	RS-232 unbalanced, RS-422, balanced
Cable impedance	RS-232, doesn't matter, RS-422, 100 Ω
Special requirements	None
Maximum distance	Depends on wire gage size
Rule-of-thumb distance	100 m (328 ft), RS-232 farther
Minimum bend radius	Doesn't matter
Connector type	Many different types, sub-D common
Connector impedance	Doesn't matter
Connector requirements	Indoor only
Other considerations	RS-422: Category 5/5e/6, no baluns
Cautions	No impedance specified for RS-232 cable

TABLE 9-24   Machine Control NEC Considerations

Unrated	No longer accepted by NEC for installs
CM, CL2, CL3	Common
CMR	Rare
CMP	Available but expensive
Other considerations	Conduit size based on cable size/fill ratio

TABLE 9-25   Machine Control Helpful Hints

1.	RS-422 can substitute for RS-232, but not vice versa
2.	Shielded/grounded connectors help reduce signal emission

Conduit and Conduit Fill

Now that you've chosen your cable and determined the number of cables end to end, you must calculate the size of each bundle. You can do this by figuring the cross-sectional area of each cable. Multiply the square of the diameter of each piece by 0.7854 (which is a variation on the πr^2 formula you learned in high-school geometry). Then add the areas together and compare the total to the three resultant columns in Table 9-26. This will give you a ballpark number for the total area. No bunch of wires will squeeze together perfectly, but it's a rough start.

THE MYSTERY OF CONDUIT PHIL

Quite a few installers are curious why the percentage of fill varies so widely across one, two, three, or more cables. It also is surprising to some that the percentage of fill does not follow that same one–two–three order. The reasons are really quite logical. First, a single cable is almost perfectly round. Even when "lumpy," it is more round than even the tightest bundle of loose cables. It also moves as one unit. These two factors mean that it has the largest percentage of fill. Two cables, in contrast, are the worst shape, because they will always be a figure-eight, which is about as far from round as one can get. Therefore, they have the lowest percentage of fill. Three cables form quite nicely into a triangle, which is a lot closer to a circle than two cables are. More than three gets even closer to a circle. So more than three is second only to a single cable in allowable fill.

TABLE 9-26   Conduit Fill

Conduit Size (in)	Equivalent Area (in²)	2 Cables 31% Fill Area	3+ Cables 40% Fill Area	1 Cable 53% Fill Area
0.5	0.30	0.09	0.12	0.16
0.75	0.53	0.16	0.21	0.28
1	0.86	0.27	0.34	0.46
1.25	1.50	0.47	0.60	0.80
1.5	2.04	0.63	0.82	1.09
2	3.36	1.04	1.34	1.78
2.5	4.79	1.48	2.34	2.54
3	7.38	2.29	3.54	3.91
3.5	9.90	3.07	4.62	5.25
4	12.72	3.94	5.90	6.74
5	20.00	6.20	8.00	10.60

It is recommended that a single cable not exceed 53 percent fill area. Two cables should not exceed 31 percent, and all other cable should not exceed 40 percent fill.

You can convert from area to diameter to determine the sizes of holes in walls, the width between walls for riser installations, and the size of core drilling (if you are putting holes through concrete walls or floors). Simply divide the total area by 0.7854 and take the square root of the result. That will be the diameter of the bundle.

If you intend to put the cable bundle in conduit, the total area can give you a rough idea of the size of conduit to install. Just remember that you cannot fill a conduit 100 percent. Most installers try to stay below 60 percent. The NEC standard is 40 percent.

You will understand the disaster that can occur when an architect is told to save money on a design. One of the first things he is likely to do is reduce the size of the conduits. If you have planned for a 2 1/2-in conduit that is changed to 2 in, your installation is in trouble. Sixty percent of a 2 1/2-in conduit is equivalent to 76 percent of a 2-in conduit, most likely an impossible pull.

Pulling in Conduit

There are a number of tools and techniques that can be used to speed a pull through conduit. One technique is to count the total degrees of bend. Most common bends are 45 or 90 degrees.

Two 90 degree bends are very difficult to pull even a short distance. If there are more bends, an intermediate pull box should be attached to the conduit; pull the cable to that location and then repull through the rest of the conduit as a separate operation.

Pull boxes also can save you when you want to modify an installation. They can be a point at which the run can be split into other directions. In some installations, intermediate boxes can be installed, where in the future you might expand into an adjoining space. Just be sure the boxes installed have sufficient "knockouts" to expand in the direction you will want to go. Once there's wire running through the box, it cannot be removed, turned, or substituted by another box.

Just remember that, if you're planning to expand, it's a good idea to calculate the length of the cable you will need for that expansion. You will also need to calculate how much additional conduit you will need and determine whether you have the space to run it. If you have only enough space for today, you won't have enough tomorrow.

Metal Underfloor Duct

Many buildings are constructed with metal ducting in the floors. It is somewhat easier to pull cable through metal ducts than through conduit. For one thing, floor ducting often has access plates at regular intervals. This allows you to pull cables in stages, thereby avoiding angled pulls as might be experienced in conduit.

The main problem with metal ducting is that it has to be placed in the floor as the concrete is poured. It takes a lot of planning to be sure the ducting will go under the areas where your equipment will be located. Once it's in, underfloor ducting cannot be added to or changed. You also must have access to pull boxes and end points, even when the floor is carpeted.

Lubrication and Pull Ropes

There are a number of devices and techniques that can aid in pulling cable through conduit. First is cable lubricant or pulling compound. This material is an odorless, yellow-green gel that is inert and nontoxic; it has no effect on cable jackets or conduit. It is very slippery, however, and its sole purpose is to allow the cable to slide more easily. If you have a number of bends or turns, you definitely should buy a few squeeze bottles of this material.

Second is a string blower, which is a souped-up vacuum cleaner. This device blows a string down a conduit. You block off the ends of the conduits into which you don't want the string to go. Only the one left open will have air flowing down it, and the string will appear at its end. Once a conduit is full of wire, it is impossible to blow in another string. Having a string in every conduit is a godsend when you need to add that one cable you forgot.

If you don't have a string blower, the next best thing is a fish tape, available from any good hardware store. The best fish tapes are self-contained units consisting of flat metal tape on a plastic storage reel. The end can be bent as a hook and fished through conduit (or walls or floors). You can attach a pull string to the tape to start the process or use the fish tape itself to do the pull. Fish tapes are not as strong as you might think, however, and you can easily break the end off as you're pulling. It's better to pull in a string.

It is a good idea to keep a string in every conduit even after the installation is finished. (You'll find out that no installation is ever finished.) Whenever you use the pull string, attach another piece of string to whatever you're pulling; that way you pull the original string back in when you're done, or you replace it with a piece of new string. If it's a short pull, tie the two ends of the string together while you're using it. That way you can never lose it because you always have a loop.

The first thing you pull with your pull string is some rope. This rope will pull your cable bundle in. Regular nylon rope is good, or you can buy steel-reinforced nylon. Whatever you buy, check its maximum pulling tension. If you have two or three burly installers, you can easily have 700 lb of pulling tension at the end of that rope!

The next item you will need is a basket puller. If you are pulling more than just a few cables, a basket puller is worth the expense. It is a mesh of wire brought to a central point, at which a ring is clamped or welded. The cable bundle is inserted into the wire mesh and a loose wire is woven around the basket, securing the bundle. Duct tape is often added to make the assembly as smooth as possible. The ring at the end makes attaching the pulling rope a breeze—although a refresher course in Boy Scout knots can also work wonders.

How to Pull

Annealed copper begins to elongate, or stretch, when the tension on it exceeds 15,000 lb/in^2. For multiple cables or multiconductor single cables, simply determine the gage of each wire inside it and add up the maximum tension for all parts concerned. In a pull of multiple cables, the pulling tension must be equally divided across all the cables. For individual wires, the values shown in Table 2-3 apply.

Just remember that you can easily rip apart an entire bundle of wires by over-zealous pulling. Be sure to have one or two people feeding the bundle at one end. If you are using plenum cable, it is especially important that they be able to see into the drop ceiling or raised floor.

Edges of lighting fixtures and mounting brackets, protruding tips of screws, and other sharp objects can rip open cables like a razor blade. If you feel a sudden resistance to your pull, make sure it's not something cutting into your cables. Keep your pulling smooth. Jerking can easily subject the cable to tension beyond the maximum allowed.

Service Loop

Whether you pull your cable with or without conduit, be sure you pull more than you need. You can always cut off the excess. Having too little wastes all your time and all the cable you just pulled. It is good wiring practice not to cut your cables until they have been dressed, that is, neatened and placed the way they will be permanently tied down. Even then, be careful. If you have a device that slides out for adjustment or routine maintenance, you will want to leave enough cable attached to allow for this extra movement. This is called a *service loop*.

GROUNDING DISCLAIMER

Following is an overview of grounding procedures and practices as they relate to audio, video, data, and other signal wiring. The ground problems considered are those that appear with low-voltage systems, such as noise, RF interference, and ground loops. The ultimate purpose of this chapter is the preservation of and adherence to the safety aspects of grounding while obtaining the maximum noise reduction.

Although electrical safety ground is discussed and analyzed, this book is not intended as a guide to safe electrical grounding procedures and practices. Electrical safety ground design and/or installation should be performed only by experienced, licensed personnel, and accepted and signed off by local building inspectors or similar officials.

Why a section on grounding?

If you talk to an installer about wire and cable, the topic often turns to a discussion of hum and noise. Although these can be wire-related problems (with wire-related solutions), they can just as often be ground-related problems. Only the correct design of a grounding system can solve a ground-caused noise problem.

GROUNDED DOG

I cannot resist telling what is probably an apocryphal story about grounding. An elderly lady called the phone company to complain of something odd that happened each time her phone rang. Just before the phone rang, her dog, who was chained in the yard, would bark insistently. Then the phone would ring. After a number of calls to "repair," the phone company finally sent someone out to investigate this obvious "crackpot." The installer checked the phone and could find nothing wrong. But as he was leaving, he heard the dog barking and then the phone rang! So he went outside to check the dog. To understand what he found, you have to realize that the phone company sends 50 volts DC to power up each phone and 90 volts AC to make it ring. Often they use ground to establish one side of the ringing circuit (while maintaining a balanced line for the talking part). They establish this ground by inserting a rod in the ground and connecting a wire to it. In the case of our prescient pooch, there was a telephone pole in the yard and he was chained to the ground post. The post had rusted over the years, however, and was no longer a very good conductor. When the 90 volts came through to ring the phone, it took the path of least resistance—the dog! Well, 90 volts is a good jolt that would start anybody yelping. Because the dog was not as good a conductor as the metal ground post should have been, the dog got a shock but the phone still did not ring.

> However, this pooch would get so upset at the number of shocks that it would lose control and urinate on the post. This wet ground suddenly became a good conductor; 90 volts went to ground—and the phone rang! Replacing the ground post (and moving Rover to a less threatening location) solved the problem.

Virtually all data and broadcast construction projects run into problems of grounding. These problems occur primarily because there is a conflict between issues of safety (grounding to prevent electrical shock to equipment users) and electronic noise reduction (using "ground" as an electronic "dump" for noise and interference). These two uses often are not compatible and can sometimes be in direct conflict with one another.

What is ground?

The definitions of ground, physical and electrical, are closely related. The ground, as in the dirt under our feet, has been known for more than 150 years to be a good conductor. Early in the history of telecommunications, with the advent of the telegraph, it became known that the earth could substitute for one of the wires needed to make a circuit. This required that the other wire be insulated from ground; it was usually hung on wooden poles, what we would call telegraph poles (or, later, telephone poles). Understanding that the ground could substitute for one of the conductors saved 50 percent of the cost of wiring and a great deal in the labor needed to install it.

Ground and lightning

Lightning is the electrical difference, or potential, between a cloud and the ground. A lightning bolt is the discharge of that potential difference from the ground to the cloud (although the ionized plasma path of air molecules makes it look like the lightning is going the other way). Lightning rods give the lightning a very low-resistance path to ground and, therefore, saves other structures, vehicles, or people from being in that conductive path. The idea of a low-resistance path to ground is one of the keys to understanding the uses of ground in electronics.

Ground and the telephone

When telephones and later audio equipment were invented (microphones, mixers, amplifiers), it was realized that some of the audio signals were so weak that the slightest electrical noise could interfere with them. To prevent outside interference, a shield of wires was woven around the signal wires. Later, metal foil was also used (see the sections on shielding). For any shield to work, it must be attached to ground. If the connection to ground is a good one, the ground will have a low resistance; the noise will prefer to flow into the ground rather than stay around the signal wires and interfere with them.

Ground and capacitors

In high-voltage, high-power transmitters, many devices, especially capacitors, store electrical charges that could be lethal to those working on the equipment. Special circuits are installed, which "bleed" away the lethal charges to ground. Sometimes a metal-ended wooden stick is provided, with a wire attached to the metal end (Figure 9-1).

The wire connected to the metal rod has a round lug attached to the other end. This must be securely attached to a good ground point. When grounded, an engineer can touch devices and surfaces inside high-voltage equipment. If some-

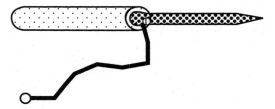

Figure 9-1   Grounding rod.

thing is highly charged, the metal tip and wire discharge the voltage to ground, thus ensuring the safety of the engineer. These rods, called "grounding sticks," are still found around broadcast transmitters and other high-voltage equipment.

A Very Short History of Power Distribution

With the rise of power distribution to homes and businesses, the lethal nature of even 120 volts (or 220 volts in Europe) became immediately apparent. It was the lethal nature of these voltages that convinced Thomas Edison that AC was the wrong choice for home distribution of power. To prove his point, he would often wire fences with 100 volts or more of alternating current. Pets that wandered into them were, naturally, electrocuted. Edison saw this as a fitting example of the lethal nature of AC, and was not above using it to frighten the public away from his principal competitor, inventor Nikola Tesla.

Edison wanted to have DC adopted as the power standard; 12 or 24 volts (such as used in motor vehicles) is a very safe voltage, and it is virtually impossible to hurt yourself with such a low voltage. There were two problems, however: resistance and distribution.

If you know Ohm's Law (see addendum), you know that, with a constant voltage (12 volts, say), the size of the wire must increase (thereby reducing resistance) to carry more current. In a toaster, for example, the wire feeding power to it should be big enough so that little is lost in the line and is delivered instead to the appliance. The power (volts times amps) could be increased on the same wire, if you increased the voltage. It was much cheaper to raise the voltage than to supply a bigger wire.

The other factor was distribution. Former Edison employee George Westinghouse left Edison because he believed alternating current was the best way to deliver power. He went to Pittsburgh where, in the 1880s, he powered electric street cars with AC.

The key to distribution is the transformer, which allows you to trade voltage for current. It consists of two coils of wire wrapped around an iron core. Because the magnetic fields of the two coils cross, the energy introduced in one coil will appear in the other. By changing the ratio of the turns of wire in each coil, you can change what comes out the other side. If it goes from few turns on the entering side (called the "primary coil") to many turns on the output (called the "secondary coil"), you can make the voltage much higher.

You can't get something for nothing, however. If you step up the voltage, the current will go down, and vice versa. If you determine the power on both sides (multiply voltage by current), you will find that they are identical (or close to identical, allowing for natural losses in the transformer). You will often hear of power in a transformer expressed in volt-amperes, partly because this power is merely transformed, that is, not being used up or doing work (and partly because of technicalities having to do with the nature of AC that are beyond the scope of this discussion).

It was this transformer that Westinghouse used to raise the voltage on his streetcar lines. Thus, a small wire could handle a tremendous amount of power. A 1,000-volt line at 1 amp was equivalent in power to a 12-volt line of Edison's at 83 amps! A wire that could handle 83 amps was a lot bigger than one that could handle 1 amp. At any place along the system, Westinghouse could put in another transformer that would go from 1,000 volts down to, say, 100 volts. Of course, the current would go from 1 amp to 10 amps, so the wire would be bigger from that point on. But it meant that power could be economically delivered along reasonably small wires (at high voltage) and then "stepped down" to a usable voltage at any point along the way.

It is no wonder that power transmission lines are thousands, even hundreds of thousands, of volts. Current (and wire size) can be kept low. The only problem is keeping that 10,000-volt line away from people, a problem that persists to this day. So Edison was right, but the almighty dollar (and technology) proved his downfall, at least on this subject.

ELECTRICAL DANGER UNDERGROUND

I vividly recall when, as a young boy, I watched workmen in the street installing a new sewer pipe. Down in the hole, the workmen had removed all the dirt around the pipe and had begun to use power saws to cut the old pipe away. They did not cut completely through the pipe, however, and left one small connection. Then one workman in rubber gloves used a special insulated saw to cut through the last piece. Another workman, also in rubber gloves and holding a meter with two leads, touched one lead to each side of the cut. It was not until years later that I understood what they were doing. The sewer pipe was metal and probably the main ground connection for all the buildings in that area. If that pipe was an especially good ground, there could be a lot of electrical energy flow-

ing through it at that point. If cutting the pipe had inter-
rupted that flow, the workmen could have been
injured or killed by touching both sides of the severed
pipe. By using a voltmeter on each side of the pipe,
they could tell how energized that section of pipe was.
As it turned out, it must have held very little electrical
energy, because they proceeded to take off their rubber gloves and cut up the pipe with regular saws.

Protecting the Customer

One way of protecting the user was to enclose plugs, switches, and wires in metal boxes or tubing, called "conduit." Once
the conduit was grounded, then any wiring fault, break,
short circuit, or other potentially lethal change would send
the electrical power to ground instead of through the user. It
is in this context, as safety grounding for electrical power
distribution, that ground is most commonly used. Like any
other use, however, the ground point chosen (the connection
into the earth) must be a good, low-resistance one. The better the ground, the safer the installation.

A simple safety ground is established by tying a wire to a
cold water pipe. Cold water pipes invariably go underground
and attach to larger, even deeper pipes. Therefore, they make
an easy connection scheme for ground. If a portion of metal
pipe between the ground wire and the ground itself is broken,
missing, or replaced by plastic pipe, there is no ground or the
ground is imperfect at best. The danger of electrical shock to
that building's occupants increases significantly.

Regardless of how the safety ground is established, even if
it is made by driving copper rods into the ground, there is one
absolutely predictable thing that will happen: No matter how
good the ground point itself is, by the time you add wire or
conduit to that point, and the farther that wire or conduit goes

from the source, the less protection you will have. The reason is, all conductors have resistance; the longer the conductor, the greater the resistance. If there is enough resistance, any electrical fault can find something other than that wire to be a good avenue to ground. If that "other path" is the user, then the safety ground scheme has not succeeded. Most electrical installers try to minimize this possibility by using large-size conduit and, sometimes, an ohmmeter to look at the resistance to ground while the building is being constructed.

It should be stressed that the quality of ground at any particular point in a building, compared to some other point in the building, can be dramatically better or worse. The quality of ground at any particular point depends on the items listed in Table 9-27.

TABLE 9-27   Conduit Installation Quality

The size or thickness of the conduit used. Thicker conduit has less resistance.

The type of the joints, elbows, and other connectors used with the conduit. Threaded or welded joints are far superior to the more common set screw joints used with lower resistance, greater reliability, and EMT conduit.

The quality of the workmanship in the conduit installation. Even the best materials can be compromised by shoddy workmanship.

The quality and reliability of the central ground point to which the building is attached.

Ground quality is a good argument to use a reputable professional electrical contractor in your installation. Rusted, corroded, or abused grounding rods, dependence on cold water pipes chosen without regard to their actual ground potential, and high-resistance connections to the building interior can render a ground system next to useless.

Ground and the NEC

Article 250 of the NEC addresses almost every kind of ground. The main thrust is personal safety, not signal purity. If you're interested in the subjects listed in Table 9-28, get a copy of the NEC (see page 349).

TABLE 9-28   NEC Sections

Systems, circuits, and equipment that are required, permitted, or not permitted to be grounded

Location of grounding connections

Types and sizes of grounding and bonding conductors and electrodes, or methods of grounding and bonding

Conditions under which guards, isolation, or insulation can be substituted for ground

Other Types of Ground

There are certain installations where running copper pipes or rods is not possible. Planes, boats, and cars are good examples. In those cases, a certain point is designated as the lowest potential point and everything to be grounded is attached to it. It is most often the negative terminal of the battery. Because there are many wires running to that point, often the metal of the vehicle is used and the negative battery terminal is attached to the body of the car. Then ground can be established by connecting to the body of the car anywhere else.

The steel used in cars (or planes or boats) is not a very good conductor, however. The resistance difference between one point and another can be significant, which would cause a voltage difference between those two points. For lights and other non-signal uses, this is a minor problem. For radios,

CD players, microprocessor-controlled subsystems, and other signal devices, this difference can generate voltage (and noise) between the ground connection of those devices and their positive battery supply connection, thus reducing their performance or rendering them useless.

Nothing will be as effective as establishing and connecting all grounds to the actual point chosen, that is, the negative pole of the battery itself. There are a number of wiring devices to allow you to do just that: make multiple connections to ground at your battery. The reduction of ground voltage and noise pickup can be dramatic. Further, because car battery voltages are low, fewer safety issues are involved.

Safety Ground and Racks

In the data and broadcast world, nearly all equipment is mounted in upright metal cabinets or racks. Doing so makes it easy to establish safety ground.

First, the rack itself might be grounded with a large wire running to a central ground point (Figure 9-2). Second, and most likely, conduit running in and out of the rack to deliver AC electrical power will provide ground for the equipment (Figure 9-3). Third, virtually all equipment has a three-pin power cord. That third round pin on the plug is connected internally to the metal box or chassis of the equipment (Figure 9-4).

The power cord connects that third pin to the third pin of the AC power receptacle and from there to a third green wire inside the conduit. The green wire usually runs to a ground point in a circuit breaker box or similar central point. Because the equipment is grounded and the equipment is mounted in a rack with metal screws, the rack is grounded (Figure 9-5).

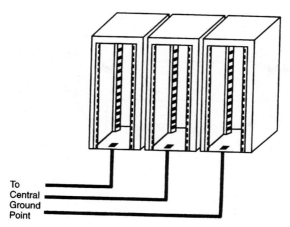

Figure 9-2 Central ground point.

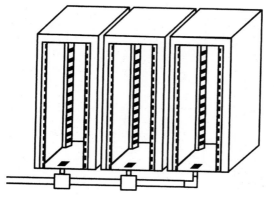

Figure 9-3 Equipment ground.

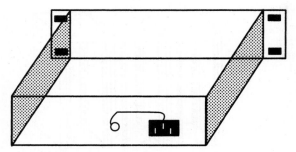

Figure 9-4 Third pin ground.

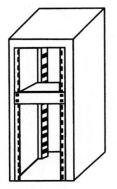

Figure 9-5 Rack ground.

And Then There Is Signal Ground

There are other ground circuits going into and out of equipment. There are the shield connections for signals entering and leaving each box in the rack. The shield is grounded to the metal box of that piece of equipment. Shielded cables are often attached to both a transmitting box and a receiving box, so there can be many paths to ground. Figure 9-6 gives you an idea of all the different ground connections possible in a group of racks.

These grounds include grounded racks, the conduit attached to the racks, the green wire in the conduit, the power cord that connects the green wire to the equipment,

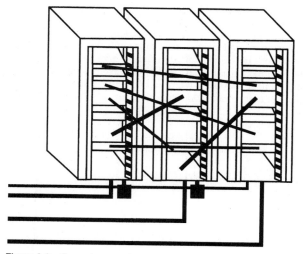

Figure 9-6   Ground connections.

and the input and output cables. The example in Figure 9-6 shows racks next to one other, but it's more likely there would be more racks in different rooms, possibly even on different floors, and the signal wiring would be in conduit, trays, or raceways running between these locations.

Ground Potential

Here, then, is the germ of the problem. An audio, video, or data cable could be attached between two pieces of equipment in separate rooms, separate floors, or even separate buildings! If that connection were better than the safety ground at both ends of the cable, or if one safety ground were much better than the other, then significant voltages, noise, and interference could travel down the shields of the signal cables.

Figure 9-7 shows the problem in schematic form. R_g is the separate ground system used to ground the racks, R_c is the resistance of the conduit from rack to rack (which also could be from room to room), R_w is the green ground wire in the conduit, and R_s is the resistance of the signal ground from one piece of equipment to another. The "arrow" of multiple paral-

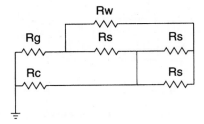

Figure 9-7   Ground potential.

lel lines is the schematic symbol for ground. The conduit ground is one path, but the signal grounds offer many parallel paths. Each of these paths has resistance (as the schematic shows), and some have more resistance than others.

Part of the confusion in understanding grounding is that there are so many ground types. It will be easier if we look at one type of ground at a time. For instance, let's look at the ground supplied by the conduit itself. We can simplify our schematic to show the conduit ground as shown in Figure 9-8.

Remember, even the conduit has resistance, and every piece of conduit adds to that resistance. To understand what effect this has, you must understand the idea of a voltage divider, which is what this circuit is.

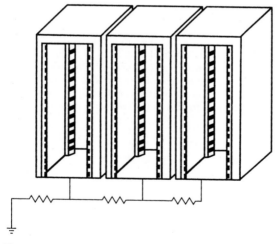

Figure 9-8 Conduit ground.

Ground connections often have voltages flowing down them. These can range from minute RFI or EMI signals, from which the conduit is shielding the wire inside, to lethal voltages from miswiring or accidental faults.

Imagine that our racks are represented by points A, B, and C in Figure 9-9. Let us assume that in rack C there is a voltage on ground for some reason, and that this voltage is therefore on the conduit. This voltage goes through three pieces of conduit (the "resistors") before getting to a central ground point. Now assume there are pieces of equipment in racks A and C that are wired together, shown in Figure 9-9 as resistor X.

It would not take much to make the signal connection between A and C better than the ground connection. You might wonder how a little copper wire could ever be better than a big, fat conduit (even if it is steel). If the conduit is rusty or badly installed, a good copper ground wire generally produces lower resistance.

The problem could be that the actual length of the conduit is many feet (maybe even back and forth to a breaker box), whereas the wire is only a few inches from rack to rack. Further, there might be more than one wire from A to C. If we were talking about patch panels and routers, there could be a hundred wires from A to C. Even with a short, perfect

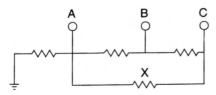

Figure 9-9   Ground loops.

this circuit, curing the ground loop problem but also lowering the potential noise-reducing effect of the shield that was disconnected.

Another quick fix is removing the third (ground) pin from the power cord of one piece of equipment or inserting a 3-to-2 pin adapter. If the right unit is chosen, this breaks the ground path because the equipment is no longer grounded. If the equipment is mounted in a rack and the rack or any other piece of equipment in the rack is grounded, then that offending piece is still grounded.

Sometimes a lazy technician will remove all the third pins and unground the entire rack to solve a grounding problem. If the specific piece of equipment is identified, it can be mounted in the rack with non-conductive washers and plastic screws to prevent any metal-to-metal contact rendering the piece ungrounded.

Although such actions solve the noise problem, removing the third pin disconnects the safety ground and presents a serious electrical shock hazard. For the liability reasons alone, this solution should be avoided.

The Fiber Optic Solution

One of the key advantages of fiber optic cable over copper cable is immunity to ground problems. Because fiber is made of glass, there is no metal contact and, therefore, no ground problem. There are very few machines that are set up to use fiber, however, and those that are usually come at a premium.

Converting to fiber means buying an extra box for each output and each input, a significant added expense. If your sole purpose is to cure a ground problem, such conversion cures only the problem for those boxes that are converted. All other equipment will be unaffected.

It is better to fix the ground problem than try to avoid the problem by going to fiber. The only time fiber may solve an insurmountable ground problem is point-to-point connections or networking between multiple buildings, such as a campus setting, where a common ground point is geographically and logistically impossible.

Medium-Quick Fixes

In the data world they avoid grounding problems altogether by using unshielded twisted pairs (UTPs). This has proved to be a wholly acceptable solution. Although the data solution is an effective and easy solution, it does not address the actual cause of the problem: poor grounding or a poorly designed ground system.

In general, digital signals are more rugged than analog and can withstand more interference and noise. Thus, converting to digital audio and digital video may help reduce noise problems. The use of UTP for digital audio and digital video is only just beginning to be recognized, and there are only a few professional, high-quality pieces of equipment where using twisted-pair data cable is an option.

The Electronics Industry Association is studying grounding, but they are between the proverbial rock and hard place. They cannot recommend that all signal cables be unshielded. In some cases, this is not possible: Hospitals, airports, and other locations with sensitive life-critical systems are very wary of interference, so they often specify shielded cable.

Telescopic Grounds

In the audio world there is a solution to ground loops called the *telescopic* ground. A telescopic ground works only with a cable that is a balanced line, that is, one that has two wires to carry the signal and a separate shield. In a telescopic ground, the shield is connected only at one end, which prevents the completion of the ground loop. It provides one path instead of two for noise and interference to go to ground, instead of two, so the noise reduction is less.

A telescopic ground works best when the end that does have the shield connected to ground is the source end, so any installer must be sure exactly which end he is hooking up. Shield effectiveness is reduced as you travel farther from the grounded end. (That's how the shape of a collapsible telescope, going from large to small, describes the shielding obtained.)

Telescopic grounds cannot be used in unbalanced circuits such as video coax because the shield is one of the two conductors necessary to send the signal. That is, the shield is both the noise-reduction portion of the cable and a signal path. Unground it and you will have the world's noisiest circuit if the signal gets through it at all. The other path, taking the place of the shield that was disconnected, will be established through some other ground path through other equipment! You might as well hang a single wire in midair as an antenna for noise and interference.

The Illegal Solutions

If one could eliminate the safety ground, the problem would go away, but so would the safety of the equipment. No building inspector in his right mind would sign off on such a practice.

Equally illegal is disconnecting the green wire in the AC receptacle. Also illegal, but less noticeable, is leaving the conduit as it is and attaching the third-pin connections from the boxes to a separate ground. This means that the safety ground is no longer the conduit but a separate circuit.

Analyzing the Problem

All of these solutions to grounding problems contain one or more of the flaws listed in Table 9-29. What you really need is a system-wide or building-wide solution.

TABLE 9-29   Grounding Problems

1.	They solve the problem for that equipment only
2.	They cost a lot of money
3.	They trade noise reduction for ground-loop elimination
4.	They compromise safety

The Real Ground Loop Solution

The key to ground is to have the same potential everywhere and a lower resistance (potential) than any other ground. (That is, if your drain is bigger, offering lower resistance, most of the water will flow down it.) The only option is to eliminate the difference in potential between any points wired with cable. If the electrical potential were the same throughout a building, then connecting any of these points to any other points would generate no voltage difference and therefore no ground loops, hum, or noise.

The aim in designing a ground system is equal potential at all places. Although this may seem impossible, especially with a building with many equipment areas, even split on many floors, it is not impossible and can be realized easily.

The Legal Solution: Star Ground

There are a number of ground schemes intended to eliminate ground loops. The most common is called a *star ground*.

In a star ground, a point is chosen as the lowest potential. There are two reasons a star ground works. First, the resistance is lower than any other ground system in the installation. Second, the length of the ground connection between a group of racks and the central ground point is always the same.

How can you get a low-resistance conductor? Get a big wire! A "big wire" can take many forms. You can separately install a big wire. The bigger the wire and the lower resistance, the more it will "overpower" every other path to ground and the better it will work. "Big wire" starts at 10 AWG and goes up from there. By the time you're up to 4/0 (0.608-in outside diameter), it will take some serious money and hefty hardware just to connect it to racks.

You can use flat-braid wire, available in a number of sizes. It has very good flexibility and is easy to connect to grounding hardware, but it is expensive. Braiding is the most expensive step in a wire-making factory.

Perhaps the most popular solution is copper strap. This is bare copper, available in a number of widths (and corresponding resistances). Being wide and thin makes it is easy to bend, turn corners, and attach to hardware. Although expensive, it is probably the cheapest solution when you compare resistance per foot versus cost.

One factor that seems to have escaped notice is the requirement for plenum ground wire. Your ground connection no doubt will travel in raised floors or drop ceilings, but I am unaware of any plenum-rated ground wire. If you use bare wire, bare braid, or bare copper strap, then there is no plastic on the wire and the metal wire automatically meets plenum requirements. Because the wire is at ground potential and carries no inherent voltage or signal, it is safe by definition and need not be jacketed.

You might not want a bare copper strap or wire running everywhere. You might wish to "jacket" or insulate it at various points. Remember, anything metal that it touches will be grounded by it. You may wish to avoid air conditioners, motors, and other devices that could introduce noise into your ground strap, thus transmitting it everywhere in your system. This is one advantage of jacketed wire, which will ground only where you strip off the jacket and want it to be grounded.

If you use bare wire or strap, just be careful how you place it. You might even want to get a copy of the building blueprint before you lay the strap and choose the most direct path with the least contact. Don't waste your noise-reduction potential on building items that don't concern your installation.

Your central ground point

The first step in installing a star ground is to run your strap or wire from the building central ground to a central ground point in your own installation. This point is most often a brass plate with multiple holes drilled in it (Figure 9-10). By having these holes tapped for large brass screws, the arm connections of the star can be established easily.

The strap from the building central ground point should be firmly attached to this block. It can be bolted or soldered

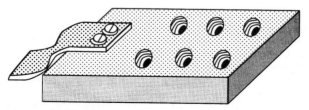

Figure 9-10   Ground plate.

to the block. Bolting gives lower resistance initially, but soldering gives a longer bond. If you are bolting the strap or wire, use fine sandpaper on the wire or strap and also on the block where the connection will be made. This reduces the corrosion or oxidization to a minimum before installation.

If you are soldering, an iron or soldering gun will be ineffective on something as massive as a strap or brass plate. You will have to use a hand torch, available at any hardware store. Be sure you have sufficient flux with the solder to allow it to flow over a surface this large. For the ultimate in connection, you might consider silver solder, which is the lowest-resistance solder commercially made.

The arms of the star

The key to a star ground is that all the ground conductors, or arms of the star, are of equal length. If the ground leads are of equal length, all the ends of the star are, by definition, at the same ground potential. Even though these arms have resistance, if the wire or strap is of an identical size and type in every arm and the distance is identical in every arm, then the resistance of every arm is identical. This means that the potential between the end of any arm and the end of any other arm is identical. But it must be at each end; you cannot connect halfway down one strap and at the end of another. Only the potential at the ends is identical. Because the potential between any two end points is the same, it doesn't matter how long the arms of the star are as long as they are the same length.

Of course, your installation and your racks are most likely not configured in the pattern of a star. They may be close together or far apart, even on different floors. As long as you keep the ground connection to each location the same length, the effect of the star ground will be maintained.

Figure 9-11 shows four racks connected in a star ground configuration. The idea that the ends of the star are at equal potential because they are equal length is critical to the success of this configuration. Length is so critical that, if you have rooms that are both near and far from the central

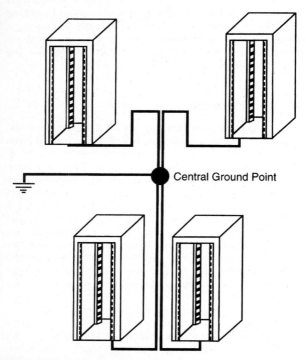

Central Ground Point

Figure 9-11   Racks connected in a star ground configuration.

ground point, the farthest point determines the length of the ground conductor to every destination.

With this configuration, a nearby room might have to have a coil of ground wire hidden under the floor or somewhere else, making up for the fact that it is not as far from the central point. Only the end of that wire is connected at the ground point. Care should be taken that the stored loops are insulated or, if bare wire, do not touch each other or any nearby conducting surfaces. Ground wires may cross the conduit, but should never touch except where each rack is grounded and at the main building ground point. If they touch, the resistance and electrical length will be different than all the other ground wires, the end points will not be at the same potential, and the noise reduction between that rack (or that room) and any other will be compromised.

What's the Standard?

NEC Article 250-21, which concerns "Objectionable Current over Grounding Connectors" in section D, clearly states that "Currents that introduce noise or data errors in electronic equipment shall not be considered the objectionable currents addressed in this section." In other words, they are concerned only with the safety aspects of grounding, not with noise, hum, ground loops, or data errors.

NEC Section 220-21 b-4 specifically allows an installer to "...take other suitable remedial action satisfactory to the authority having jurisdiction." This means you have to convince only the building inspector or similar authority that your additional ground strap or wire does not compromise the safety aspect of the building ground system. You should be able to make a good case that the additional grounding actually increases the safety of the installation.

Wire gage for grounding

In order that the ends of the star are at very low potential (low resistance), they must be large-gage wire: 10, 8, or 6 AWG is probably the minimum. The AWG number should get smaller (and the wire size get bigger) as the points of the star become more distant from each other. The larger the wire, the lower the resistance and the better the system will work. Wire gage generally is determined less by performance, and more by available space, flexibility, and cost. Just know that anything smaller than 10 AWG will probably give you little improvement over the existing conduit installation. You need to get the majority of signal ground conductivity over your star ground wiring.

Audio snakes and grounding

Some of today's snake cables provide not only a drain wire with every pair but also an overall foil and drain wire with the entire group of pairs. In some installations this addition can be extremely valuable. For instance, if your snake cable is going to a distant location, where running copper ground strap or large-gage ground wire is difficult or impossible, the overall foil and drain can be used to establish a "remote" ground potential. Just remember the basic rules outlined in Table 9-30.

TABLE 9-30   Ground Rules

1. Because this arm of your star is a different length (and resistance) than any other, its purpose is to ground only those pairs within the snake cable itself. If there is a remote junction box (a press box in a stadium, for instance), such a ground can be used quite effectively.

continued on next page

TABLE 9-30   Ground Rules (continued)

2.	If there are signals from other places coming into the remote location where the snake cable terminates, you must supply a star ground arm. The snake shield will not help you.
3.	The ground inside a snake cable cannot possibly be as good as a ground strap or even as good as a piece of conduit. If there are hum and noise problems from that location, a star ground arm must be installed.

Copper strap

The use of copper strap or tape is very common. This is flat copper tape, often 2 or 3 in wide and a $1/_2$ in thick. Strap can be purchased from hardware stores, plumbing supply stores, electrical wholesalers, or copper-and-brass distributors. Remember that all connections to the star should be the same length. The strap is bare, so you probably will have to figure out how to store the strap when you have extra. Bare copper can be stored in a plenum (drop ceiling or raised floor). Jacketed cable in the same installation would have to have a plenum-rated jacket to be legal.

Determining the best ground point

In each room, care should be taken to choose the absolutely best ground point. The standard is to ground each rack. If racks are bolted together, you can ground the group. It is a good idea to make sure (with an ohmmeter) that there is very low resistance between the ground point and the far-thest rack. If you can read anything (even a couple of ohms), it's probably a good idea to ground the rack group at two different points. Just remember, you must run a separate leg off the star, not just extend the strap. In only the most critical situations should you consider grounding individual pieces of equipment.

Your ground and conduit ground

Remember that even though the conduit is safety ground, you are establishing a second, even better, ground. You are not compromising the safety aspects of the conduit. If you use smaller-gage wire or small copper strap, it is possible that the conduit running throughout the installation is of even lower resistance. In that case, the conduit establishes ground, with different ground potentials in each room (and each rack), resulting in noise, hum, and ground loop problems. As noted previously, the key is to overwhelm the resistance of the existing conduit with a dramatically better (lower-resistance) path to the building central ground point.

Critical ground points

In especially critical installations that are experiencing ground problems, you can look at the video, listen to the audio, or look at eye patterns or other error-rate data indicators, and move a ground strap around until the image, sound, or data clears up. Wherever you're touching should be the key ground endpoint for that room or group of equipment.

Just to be safe, measure the potential between that strap or ground wire and the rack or piece of equipment before you test it. If there is significant voltage, you must ground that strap or wire to the rack and then run another, separate piece of strap or wire from that point in the rack to the equipment of concern. That second strap should be long enough to touch every piece of equipment you consider sensitive. The majority of the voltage will be flowing through the grounded rack and strap. Your added path will show only the fine variations in placement of ground for each piece of equipment.

When you are installing the ground strap or wire and your voltmeter finds a significant voltage difference between safety ground and the signal ground you are touching, all installed equipment should be wired, connected, and plugged in—but turned off. Large voltage transients can permanently damage equipment.

Where and how to connect the ground

Most installations bring the ground strap or wire to each group of racks. If the racks are attached together (bolted, for example), they can be considered one unit and have one ground strap. If you've installed more than five racks that have critical circuits in them, even though they're all joined together, you might want to consider a second ground strap.

Ohm's Law

Power lost in cable due to changes in resistance can be determined by Ohm's Law and Watt's Law, one of the simplest and most basic set of formulas in electronics. Here are the three formulas in Ohm's Law. They show the relationship between voltage in volts (E), amperage in amperes (I), and resistance in ohms (R):

$$E = I \times R \qquad R = \frac{E}{I} \qquad I = \frac{E}{R}$$

If any two of those numbers are known, the third can be calculated. Watt's Law throws in a fourth parameter, power (P), measured in watts:

$$P = I \times E \qquad P = I^2 R \qquad P = \frac{E^2}{R}$$

Just remember that any electrically driven device has a appropriate resistive load on the line. A bad connector or joint adds resistance, which is in addition to the normal load resistance. Therefore, the voltage divides between the correct load and the added load of the bad connector or joint, forming what is appropriately called a *voltage divider*. As the bad resistance increases over time (often exacerbated by the heat generated by the resistance at the fault), the division of voltage also increases and the amount of power lost becomes greater and greater until the line melts, a fuse blows, or the piece of equipment driving the line fails. Any seasoned broadcast engineer would proudly show a piece of melted/exploded transmission line caused by this problem. Good connectors and good installation practices keep this occurrence to a minimum.

Many house fires are caused by poor electrical joints overheating. This can either be poor connections, underrated wire, or improper wiring practices. Poor connections can include wire not twisted correctly around screw terminals or fully inserted into push-fit terminals.

Underrated wire can be something like the proverbial extension cord on the Christmas tree. It has a hundred lights on it and they are all plugged into one thin extension cord. When the extension cord melts and short circuits, everyone is surprised.

Improper wiring practices can include mixing metals, such as using aluminum house wire with copper fixtures. Even though they are perfectly installed, the two metals have a different coefficient of expansion when heated. That means that they heat up together but cool down differently, leaving an ever-increasing gap (and an increasing resistance) every time the wire carries electricity. Eventually that gap becomes such a high resistance that the amount of heat it generates melts the fixture and causes a short circuit.

All of these errors result in increased resistance and increased resistance results in overheated joints. It is not the copper that melts. (Copper melts at over 2000°F.) It is the plastic and other materials that melt first, allowing the conductors to touch and short circuit; the resulting heat caused combustion and a fire.

Appropriate gage size is dependent, therefore, on the melting point of the jacket material. PVC compounds (generally around 60°C) can be made to go as high as 105°C. Other plastics do even better, with Teflon (at 200°C) being the best.

Capacitive Reactance

$$\mathrm{Xc} = \frac{1}{2\,\pi\,\mathrm{FC}}$$

Where...
   Xc is the capacitive reactance in ohms.
   π (pi) is approximately 3.14159
   F is the frequency in Hertz (Hz)
   C is the capacitance in farads

Inductive Reactance

$$\mathrm{X_L} = 2\,\pi\,\mathrm{FL}$$

Where...
   $\mathrm{X_L}$ is the inductive reactance in ohms
   π (pi) is approximately 3.14159
   F is the frequency in Hertz (Hz)
   L is the inductance in Henrys (Hy)

Impedance Formulas

Coaxial cable

$$\mathrm{Z_o} = \frac{138}{\sqrt{\mathrm{E}}}\,\log_{10}\frac{\mathrm{D}}{\mathrm{d}}$$

Where...
   Zo is the impedance of the cable
   E is the dielectric constant of the medium between the
conductors
   D is the diameter of the outer conductor (shield)
   d is the diameter of the inner conductor

Unshielded twisted-pair cable

$$Zo = \frac{276}{\sqrt{DC}} \log 2 \left(\frac{h}{d}\right)$$

Where...
  Zo is characteristic impedance in ohms
  DC is the dielectric constant
  d is the conductor size
  h is the distance between conductors

Shielded twisted-pair cable

$$Zo = \frac{276}{\sqrt{DC}} \log \frac{(2h)}{d} \cdot \frac{1 - (h/D)^2}{1 + (h/D)^2}$$

Where...
  Zo is characteristic impedance in ohms
  DC is the dielectric constant
  d is the conductor size
  h is the distance between conductors
  D is the shield diameter

Note: Boldface numbers indicate illustrations.